Green Energy and Technology

Climate change, environmental impact and the limited natural resources urge scientific research and novel technical solutions. The monograph series Green Energy and Technology serves as a publishing platform for scientific and technological approaches to “green”—i.e. environmentally friendly and sustainable—technologies. While a focus lies on energy and power supply, it also covers “green” solutions in industrial engineering and engineering design. Green Energy and Technology addresses researchers, advanced students, technical consultants as well as decision makers in industries and politics. Hence, the level of presentation spans from instructional to highly technical.

Indexed in Scopus.

Indexed in Ei Compendex.

Benjamin Duraković

Advancing Zero Energy Buildings

Pathways to Sustainable Development and Global Impact

Springer

Benjamin Duraković
Faculty of Engineering and Natural Sciences
International University of Sarajevo
Sarajevo, Bosnia and Herzegovina

ISSN 1865-3529 ISSN 1865-3537 (electronic)
Green Energy and Technology
ISBN 978-3-032-16600-5 ISBN 978-3-032-16601-2 (eBook)
https://doi.org/10.1007/978-3-032-16601-2

This Springer imprint is published by the registered company Springer Nature Switzerland AG
The registered company address is: Gewerbestrasse 11, 6330 Cham, Switzerland

Preface

This book is intended to provide the latest research as a contribution to the growing global discussion on energy efficiency and sustainability within the built environment. The concept of zero-energy buildings (ZEB) has evolved as a solution to a growing need towards creating a more sustainable future. Although this topic has been discussed extensively in literature, there is no systematic review that integrates all significant issues of research findings, design methodology, and real-world applications into a coherent work.

The goal of this book is to provide a comprehensive understanding of the principles, technologies, and practices that support the development and implementation of ZEBs, as well as an introduction to the challenges that stand in the way of achieving those goals. Particularly, the purpose of the book is to support the global transition to zero energy buildings. This is an attempt to bridge the gap between theory and practice by integrating research findings and real-world examples with the aim of demonstrating how to integrate energy efficiency and renewable energy systems within the built environment.

This work is the result of many years of teaching and research in energy management and sustainability. It reflects synthesis of published research, case studies and personal effort to present recent development in the process of transition to ZEB standards. The book is intended to serve as a resource for students, researchers and professionals dealing with sustainability in architecture and relevant engineering.

The preparation of this book has been greatly enriched by discussions with colleagues and friends. I express my sincere appreciation and gratitude to those who contributed with their suggestions and technical advice. I would also like to thank my institution, the International University of Sarajevo, for providing an academic environment that encourages research. I also express my sincere gratitude to my family for their patience, encouragement, and support during this journey in challenging moments.

Sarajevo, Bosnia and Herzegovina

Benjamin Duraković, Ph.D.
bdurakovic@ius.edu.ba

Competing Interests The author has no competing interests to declare that are relevant to the content of this manuscript.

Contents

Chapter 1
Importance of Zero Energy Buildings in Sustainable Development

Abbreviations

CO_2	Carbon Dioxide
EPBD	Energy Performance of Buildings Directive (EU)
EU	European Union
HVAC	Heating, Ventilation, and Air Conditioning
IEA	International Energy Agency
PV	Photovoltaic (for solar energy systems)
SDGs	Sustainable Development Goals
U.S.	United States
ZEB	Zero Energy Building

1.1 Introduction

Fossil fuels remain the primary source of energy worldwide, accounting for 81% [1] of global energy supply [2, 3]. This percentage has been consistent over the past 50 years [2]. These fossil fuels are major source greenhouse gases such as carbon dioxide [4, 5]. Climate change cause intense changes in climate such as average air temperature increase resulting in the loss of ice on the south and north pole, increase of floodings, hurricanes etc. [2].

Taking into consideration all above mentioned, it is very understandable that the movement provided by the UN called Sustainable Development Goals (SDGs) 2030 has been created. The 2030 agenda is a call for action in order to make the world a better place. That includes "No Poverty, Zero Hunger, Good Health and well-being, Quality education, Gender equality, Clean Water and Sanitation, Affordable and clean energy, Decent work and Economic growth, Industry, Innovation and Infrastructure, Reduced Inequalities, Sustainable cities and communities, Responsible Consumption

B. Duraković, *Advancing Zero Energy Buildings*, Green Energy and Technology,
https://doi.org/10.1007/978-3-032-16601-2_1

and production, Climate action, Life below water, Life on land, Peace, Justice and Strong Institutions, and Partnerships for the goals" [6, 7]. As all professionals in their field are working on an SDG related to their own field, architects and engineers work on the goals of sustainable cities and communities as well as climate change indirectly [6], which leads us to the phenomenon of Zero Energy Buildings (ZEBs).

ZEBs are highly energy-efficient structures where the entire annual primary energy demand is met through a combination of on-site and nearby renewable energy sources, renewable energy communities, efficient district heating and cooling systems, or other carbon-neutral energy sources. These buildings aim to operate without reliance on fossil fuels, ensuring minimal environmental impact and supporting sustainability goals [8–12]. In order to create this balance, it is necessary to generate energy on the site using solar panels or wind turbines, or any other kind of renewable energy source to minimize energy waste and as well using clean energy [13–16].

The focus of this chapter is to show the general and in-dept concept, history and evolution of these zero energy buildings, as well as their role in the sustainable development and sustainable cities, which were mentioned. In addition, some policies and market trends that have pressured presence of ZEBs will be discussed.

1.2 ZEB Concept and Definition

ZEBs are designed to achieve net-zero energy consumption primarily through a combination of energy-efficient design, renewable energy integration, and intelligent energy management. This involves energy demand reduction via passive design, using efficient systems for heating, cooling, lighting, appliances and efficient building envelopes, as well as clean energy generation on-site or nearby.

According to the European Parliament, Zero Energy Buildings still do not have a complete definition. Although the definition has not yet been established, general suggestion is a building that produces enough renewable energy to meet its own annual energy consumption requirements. This way the use of non-renewable energy in building sector is reduced [17]. Another definition comes from Norwegian Research Centre, and it says, ZEB is a building that during its life cycle produces as much renewable energy as greenhouse gases and includes 5 phases [18]. These two definitions differ in one key respect; the first definition states that buildings produce 0 $kgCO_2eq/m^2$ of greenhouse gas emissions, and that all the energy required is produced from renewable sources. While on the other hand, the second definition implies that the greenhouse gases produced are equal to the energy made from renewable sources. However, in summary, both definitions require the elimination of operational greenhouse gas emissions.

A crucial step toward ZEB definitions is setting system boundaries, which represent the interface between the building and its environment. This implies the determination of the life stages of the building that affects the calculation of greenhouse gas emissions [18]. Figure 1.1 represents relevant terminology withing building system boundary for the NZEB balance concept. The strategy is based on Sartori et al. [19].

Weighting system is a method used to assign relative importance to different components or factors within energy generation and consumption. In addition, the weighting factors may reflect engineering considerations as well as political preferences. It helps to prioritize and balance different sources of energy and helps in evaluation of building energy performance. On the *energy consumption* side, weights can be assigned to energy-consuming activities including heating, cooling, lighting, and appliances. These weights are assigned based on the energy consumption and efficiency of a particular energy-consuming activity to overall energy demand. On the *energy generation* side, the weights can be assigned to various energy sources including solar panels, wind turbines, and geothermal systems. These weights are assigned based on the contribution of a particular energy carrier to overall energy production. In this regard, terminology used in Fig. 1.1 has the following meanings:

- *Weighted demand* is the total energy consumption of a building, adjusted by weights assigned to different energy-consuming activities or systems within the building. This helps to prioritize the various components of energy demand based on their significance.
- *Weighted supply* is the total energy generated by the building from various sources, adjusted by weights assigned to different energy generation carriers. This represents the contribution of each energy source to the overall energy supply.
- *Load* is building energy demand per each energy carrier in (kWh/y) or (kWh/m^2y). Due to its own on-site energy generation and consumption, the load may not match the delivered energy.
- *Generation* means on-site energy generation from renewable sources per each energy carrier in (kWh/y) or (kWh/m^2y). Due to own energy consumption, on-site generated energy from renewables may not match to the exported energy.

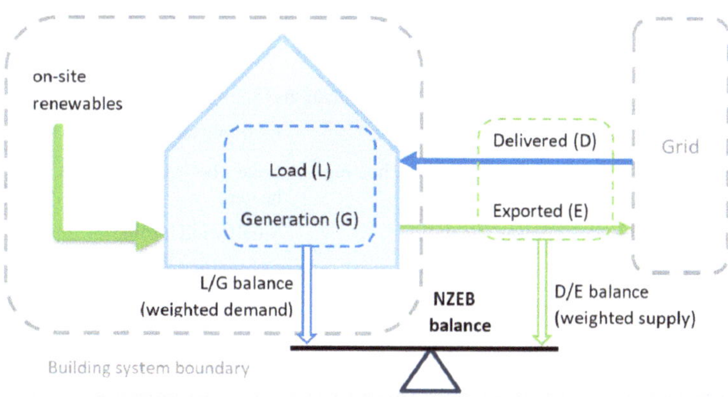

Fig. 1.1 ZEB balance in terms of load/generation or delivered/exported energy exchanges

- *Delivered energy* is the energy flow from the grid to the building per each energy carrier in (kWh/y) or (kWh/m^2y).
- *Exported* is the energy flow from the building to the grid per each energy carrier in (kWh/y) or (kWh/m^2y).

ZEBs as self-sustaining structures are designed to generate an equivalent amount of energy to their use on yearly basis, using renewable energy sources [15, 20–22]. In Table 1.1 it is shown Renewable energy sources and their applications within these buildings.

Cutting-edge technology and intricate design enable ZEBs to attain equilibrium [24]. These structures optimize energy efficiency via superior insulation, energy-efficient windows, and sophisticated HVAC systems [25–27]. Real-time energy management using smart building technology, commonly interconnected via the Internet of Things (IoT), promotes energy awareness [28–30]. Heating, cooling, lighting, and appliance functionality are regulated by renewable energy sourced from solar photovoltaic (PV) panels, wind turbines, or geothermal systems [21, 31–33].

According to the European Commission, all new residential and non-residential buildings must achieve zero emissions by 1 January 2028 for public body-owned structures and by 1 January 2030 for all other new buildings, with some exclusions [11, 34]. All organizations are introducing green buildings to decrease consumption, which harms the environment and is restricted in the future [11]. ZEBs energy efficient construction materials such as composites, windows and superior insulation reduce heat transmission and change how we think about indoor-outdoor interaction [35, 36]. These constructions become energy suppliers by producing renewable energy on-site [35]. Therefore, the ZEB-required design process shift is quite intriguing [35] since the linear design process has become complex, architects, engineers, and energy experts must collaborate from the start [35]. The ZEBs have evolved from basic structures to complex and flexible systems that govern their environment [37] via energy management system [38, 39]. This expands the global sustainability initiatives and reshapes the future of architecture and the built environment [40].

Table 1.1 Renewable energy sources and their applications [8, 23]

Renewable energy source	Application in ZEBs	Example technologies
Solar energy	Power generation, heating	Solar photovoltaic panels, solar water heaters
Wind energy	Power generation	Small wind turbines
Geothermal energy	Heating and cooling systems	Geothermal heat pumps
Biomass energy	Heating, power generation	Biomass boilers

1.3 Historical Evolution

The construction industry's perspective about energy efficiency and sustainability is transformed by ZEB design [41], as it presents a historic big step towards greener solutions and enhanced use of renewable energy sources. Energy crisis in 1970s prompted architects and engineers to rethink how to use energy in buildings [38], sparking a wave of new ideas that were directed into passive solar design [41, 42]. This enhanced research in air movement, heat, and natural light, where initial methodologies included double-glazed windows, sophisticated insulation, and architectural design to reduce the need on use of electricity and artificial heating and cooling [41]. This introduced new ideas that pave the way for ZEB designs and show ways in which energy use can be cut down through smart building design. The passive home standard, introduced in Germany in 1990, marked a significant advancement by emphasizing the construction of highly energy-efficient structures via meticulous attention to the building envelope and air quality [20, 43]. While passive homes are not directly related to ZEBs, they have provided a significant contribution to our knowledge about how buildings can be made more energy efficient [41].

Studies from 1990s and early 2000s tried to render more energy efficient building methods [41]. The Department of Energy in the U.S. initiative, The Building America, formulated regulations that enhanced the energy efficiency of dwellings [41]. The European directives including The Energy Performance of Buildings Directive (EPBD), introduced increasingly rigorous energy efficiency standards across member states [44, 45].

The rapid development in green energy technology was set at the beginning of the twenty-first century, increasing the importance of the ZEB movement [41]. During this period, the solar PV systems became increasingly affordable and efficient, that resulted in greater adoptions [27, 31, 41, 46].

Storage technologies started to be designed using innovative energy that enabled the retention of surplus energy harvested during the day for use at night or during peak demand periods [23, 31].

However, the addition of smart technology to building systems set the big stride forward for ZEBs. Building management systems and IoT gadgets enabled real-time tracking and improvement of energy use [29, 30, 47]. ZEBs became even more efficient and attractive to both homeowners and investors once they got smart heaters, automatic lights, and tools to help in planning ahead for upkeep [47].

Due to heightened awareness of climate change and the need for sustainable development over the last two decades, ZEB systems gained significant traction globally [2], strengthen with the globally adopted legislations and incentives to further to approach. The International Energy Agency (IEA) and other international organizations provided research and guidelines that have helped many to take great leaps into using ZEB [11]. The collaborative works of countries, companies, and academia the ZEB technologies and methodologies grown and proliferated further [23]. In Fig. 1.2, is a summary of historical events and developments that have shaped ZEBs today.

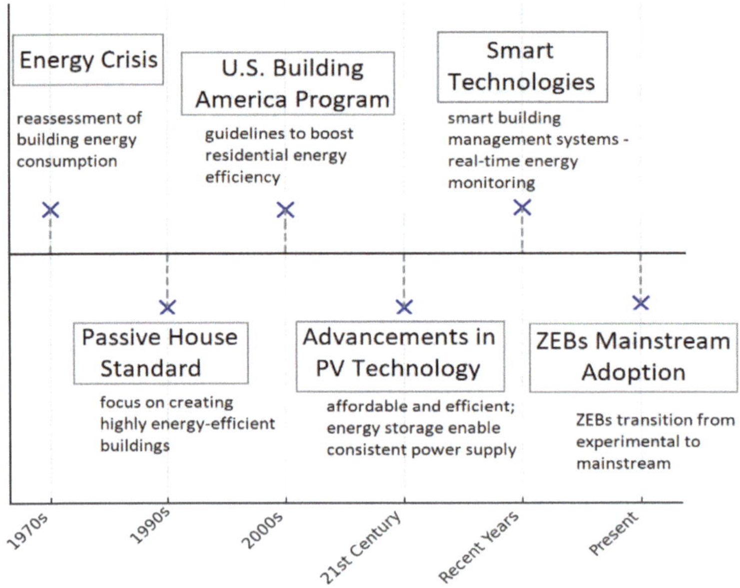

Fig. 1.2 Historical developments of ZEBs

From the small-scale experiments, in the recent decades the ZEBs have grown to become mainstream as the only way to reduce carbon emissions and ensure that at least our own homes will be powered [48]. Building-integrated photovoltaics, known as the melding of solar cells with building materials and new energy-planning tools are further pushing what ZEBs can do [48]. The development of the ZEB showcases a wider trend within the building industry to "become greener". As the technology continues to advance, ZEBs are going to become vital in creating the future built environment, following the growing consciousness among people worldwide for environmental protection.

1.4 ZEBs Importance in Sustainable Development

ZEBs play a crucial role in tackling today's global concerns regarding climate change, resource depletion, and widespread urbanization. Their contribution in the reduction of greenhouse gas emissions helps in conserving non-renewable energy resources and contributing to sustainable development. Since the energy consumption by buildings usually accounts for nearly 40% of the total consumption in advanced countries [49, 50], and construction sector account for about 39% of process-related CO_2, it was recognized as an opportunity. Therefore, the adoption of ZEB becomes a proactive action to mitigate climate change, especially they rely on efficient energy design and renewable energy systems [51–53]. In addition, energy security is considered as a one of the key challenges of today's global society, especially countries reliant on reliant on imported fossil fuels [54–56]. In this regard, ZEB would help to address these problems by generating renewable energy locally and decreasing dependence on foreign energy sources [54]. This self-sufficiency underpins energy independence at the individual building level and national level.

Through optimized building design, use of high-performance materials, and smart energy systems to minimize energy waste, energy efficiency represents the basis on which ZEBs are introduced [55]. It enhances resource conservation, and reduces running costs, therefore making sustainable living usable and economically viable [47]. Therefore, ZEBs contribute directly to some of the United Nations SDGs [6], including:

- Goal 7: Affordable and Clean Energy: ZEBs develop novelty solutions for clean energy generation and extend access to affordable, reliable, clean, and sustainable energy.
- Goal 11: Sustainable Cities and Communities: Zero Energy Buildings advocate sustainable urban development by minimizing environmental effect and improving people's quality of life.
- Goal 13: Climate Action: The wide diffusion of ZEBs creates one of the foundations for the global challenge of combating climate change and its impacts.

Besides the environmental contributions, huge economic advantages in ZEBs are recognized. Lower energy consumption converts to lower utility bills from occupants, while incentives and subsidies on the installations of renewable energy reduce their costs even further. Thereafter, operating costs normally remain lower, hence economic attractions for both home buyers and developers due to the overweighting benefits of reduced running costs [47]. Socially, ZEBs offer a better quality of life by creating healthier indoor environments. Better ventilation and lighting from natural sources raise the quality of air for the occupants' benefit. Besides, this policy emphasis on local generation creates community involvement in concerns for sustainable practices in building a culture that values environmental responsibility [47]. Table 1.2 presents the importance and benefits of ZEBs in sustainable development based on literature.

As rapid urbanization demands sustainable housing and infrastructure, therefore ZEBs become essential in this process and achieve sustainable urban development [64], serving as a driver for innovation that integrates environmental, economic, and social priorities into a more sustainable future [64].

1.4.1 Environmental, Economic, and Social Benefits

ZEBs foster building development beyond energy savings with *environmental*, *economic*, and *social* benefits. *Environmental* benefits may be ZEBs' strongest side. By generating as much electricity as they require, ZEBs cut fossil fuel and greenhouse gas emissions [5], which is crucial in fighting climate change. The use of

Table 1.2 Importance and benefit of ZEBs in sustainable development

Sections	Key points	References
Importance in sustainable development	– Climate Change: ZEBs contribute to climate change mitigation by decreasing carbon emissions and reliance on fossil fuels. Facilitates international climate objectives (Paris Agreement) – Energy Security: ZEBs reduce reliance on imported energy, enhancing energy resilience – SDGs: Contributes to SDGs 7, 11, and 13	[2] [54] [6, 57]
Environmental benefits	– ZEBs generate their own energy (wind, solar, geothermal), reducing fossil fuel emissions – Reduces energy pollution and conserves resources – Promotes sustainability through efficient design, insulation, and advanced technologies (e.g., energy-efficient windows)	[27] [58] [13]
Economic benefits	– ZEBs reduce energy costs through on-site energy generation – Higher property value due to energy-efficient technologies – Economic benefits from reduced operating costs and green building certifications – Job creation in green technology, renewable energy, and sustainable construction sectors	[47]
Social benefits	– Improved living/working environments due to better natural light and air quality – Enhances comfort, well-being, and productivity – Raises community awareness of sustainability – Promotes social responsibility and eco-friendly habits among residents	[59, 60]
Policies and regulations	– Policies: Tax credits, grants, rebates to incentivize renewable energy adoption; specific building codes (e.g., EPBD in Europe, Title 24 in California) – Energy Efficiency Regulations: Performance-based codes driving innovation in energy-efficient building design – Market Trends: Growing demand for sustainable, energy-efficient buildings as consumers and businesses adopt eco-friendly values	[61] [62] [63]
Market trends	– Shift toward sustainability in housing and corporate choices – Technological advances make renewable energy and energy-efficient systems more affordable and accessible – Growing demand for energy independence drives ZEB adoption – Policies and regulations continue to push the market forward	[63]

renewables such as wind, solar, and geothermal use of energy [65], contributes to energy conservation and mitigate climate change. ZEB energy waste is reduced via efficient architecture and advanced technologies [66], which is good way to promote built environment sustainability [54, 58, 67].

Economic benefits: ZEBs cost more to build, but they pay off in the long run. Power costs are lower since these buildings produce most of their energy on-site. Households and businesses may benefit from ZEBs since energy savings may offset initial costs. Sustainable living is growing, and customers are willing to pay extra for energy-efficient technology and green building certifications. Building ZEB laws by cities and governments may stimulate market growth. ZEBs provide green technologies, renewable energy, and building jobs. As more builders and developers use ZEB principles to boost local economies and generate sustainable industrial employment, trained personnel are needed.

Social benefits: Resident social life can be improved since ZEBs uses natural light and higher air quality via strategically placed windows, making them better living and working environments. These features promote workplace productivity and family happiness by improving occupant comfort, well-being and community awareness. Culture changes drive communities to value sustainability beyond buildings [59, 60].

1.4.2 Key Drivers

Supportive *policies*, tough *regulations*, and dynamic *market trends* have significantly impacted the development of Zero Energy Buildings. Comprehending the driving factors behind ZEBs entails understanding the rationale for their increasing predominance in contemporary architecture.

Policies by almost all governments around the world play a major role in promoting ZEBs [61]. Most countries have embraced an urgent call to action with the development of initiatives that stimulate sustainability in building. Examples of these financial incentives include tax credits, grants, and rebates associated with the installation of renewable energy systems to alleviate the elevated initial expenses of creating ZEBs [61]. Furthermore, several government programs establish clear objectives for energy efficiency; for example, all new constructions must comply with zero-emission criteria by certain deadlines [61]. The European Union's EPBD dictates that all new buildings achieve practically zero-energy status, while the California's Title 24 building code in the United States stipulates that all new residential constructions must comply with net-zero energy criteria [11, 68]. These all serve to illustrate how local policy can spur ZEB adoption. These building codes are important in developing a substantial framework that architects and builders can operate within to create a more sustainable building culture.

Regulations are another critical driver in the adoption of ZEBs is the establishment of building codes and energy performance standards [62]. In most regions, regulatory frameworks lay down minimum energy efficiency requirements for new constructions, hence pushing developers to innovate and integrate renewable technologies [36, 62]. Such regulations aim at reduced energy consumption and carbon emission, hence creating a playing field where energy-efficient buildings can thrive. Unlike prescribing the design approach, performance-based rules connect energy efficiency to certain results [62]. This will enable builders to be inventive in methods to reach compliance, therefore producing perhaps better, more useful designs that satisfy not just energy demands but also aesthetic and functional ones. Laws will guarantee that ZEBs take front stage in the building sector for a sustainable future [62].

Market trends over the last decade in terms of consumer tastes for sustainability and efficiency as well as in housing have drastically changed [69]. People want houses and businesses to represent their beliefs; they are more educated about environmental issues like climate change. This has made ZEBs even more enticing to builders and developers trying to leverage that changing market as demand has grown [69]. Besides, the developments in technology have made systems of renewable energy-solar panels and energy-efficient appliances, for example-easier to access and more affordable than ever. As these technologies continue to improve, the cost-effectiveness of ZEBs will further increase, thus luring even more stakeholders to invest in sustainable construction [63, 70]. The emphasis on energy independence fuels the trend toward ZEBs as consumers seek to minimize their exposure to fluctuating energy prices [70]. In essence, supporting policies, strict rules, and shifting market trends have been and will always be major forces in building and including ZEBs within the construction sector [63]. ZEBs will be more suited to occupy their natural place within our built environment and help to create a cleaner, more efficient future as more governments and consumers embrace sustainability [71].

1.5 Concluding Remarks

Zero energy buildings evolved from passive solar concepts in 1970s to technologically advanced smart, and self-sustaining structures. Even though passive structures are not directly related to ZEBs, they have significantly contributed to our knowledge of how toimprove building energy efficiency. They are considered as a foundation of sustainable development, which integrate advanced energy-efficient design, renewable energy generation, and smart management systems. Current definition of ZEBs vary across regions but all emphasize the use of renewable energy and mitigation of greenhouse gas emissions. Their contribution to sustainability is reflected through minimization of greenhouse gas emissions, reduction of reliance on fossil fuels, and support in mitigation of climate changes. Economic benefits include long-term energy costs reduction, property value increase, and foster green innovation and job creation. Social benefits include improved occupant comfort, well-being and foster community awareness of sustainable practices. As ZEBs are supported by strong global and national policies, and regulations, they are positioned as a key instrument in the transformation of built sectors to sustainable cities and communities worldwide.

References

1. Heffron RJ (2022) Achieving a just transition to a low-carbon economy, pp 1–162. https://doi.org/10.1007/978-3-030-89460-3
2. Olabi AG, Abdelkareem MA (2022) Renewable energy and climate change. Renew Sustain Energy Rev 158. https://doi.org/10.1016/j.rser.2022.112111
3. Heubaum H, Biermann F (2015) Integrating global energy and climate governance: the changing role of the International Energy Agency. Energy Policy 87:229–239. https://doi.org/10.1016/j.enpol.2015.09.009
4. Geography bytes from all over the globe: February 2008. https://rashidfaridi.blogspot.com/2008_02_01_archive.html. Accessed 12 Jan 2025
5. Stephan A, Stephan L (2020) Achieving net zero life cycle primary energy and greenhouse gas emissions apartment buildings in a Mediterranean climate. Appl Energy 280. https://doi.org/10.1016/j.apenergy.2020.115932
6. Sustainable development goals. United Nations Development Programme. https://www.undp.org/sustainable-development-goals. Accessed 27 Dec 2024
7. Sustainability-linked bonds will be eligible as European Central Bank collateral. Insights | Sidley. https://www.sidley.com/zh-hans/insights/newsupdates/2020/09/sustainability-linked-bonds-will-be-eligible-as-european-central-bank-collateral. Accessed 12 Jan 2025
8. Maduta C, D'Agostino D, Tsemekidi-Tzeiranaki S, Castellazzi L (2025) From nearly zero-energy buildings (NZEBs) to zero-emission buildings (ZEBs): current status and future perspectives. Energy Build 328:115133. https://doi.org/10.1016/J.ENBUILD.2024.115133
9. Ma Z, Arıcı M, Sun Y, Singh S, Shahsavar A (2024) Towards net-zero energy/emission buildings for sustainable development. Energy Sustain Dev 80. https://doi.org/10.1016/j.esd.2024.101448
10. Nearly-zero energy and zero-emission buildings. https://energy.ec.europa.eu/topics/energy-efficiency/energy-efficient-buildings/nearly-zero-energy-and-zero-emission-buildings_en. Accessed 27 Dec 2024
11. Maduta C, D'Agostino D (2024) Readiness of zero-emission buildings (ZEBs) implementation in the European Union. E3S Web Conf 523:04009. https://doi.org/10.1051/E3SCONF/202452304009
12. Bahrami MR. Zero energy building: systematic literature review (6376). https://doi.org/10.31224/OSF.IO/FSCG2
13. Lou HL, Hsieh SH (2024) Towards zero: a review on strategies in achieving net-zero-energy and net-zero-carbon buildings. Sustainability (Switzerland) 16. https://doi.org/10.3390/SU16114735
14. Hernandez P, Kenny P (2010) From net energy to zero energy buildings: defining life cycle zero energy buildings (LC-ZEB). Energy Build 42:815–821. https://doi.org/10.1016/j.enbuild.2009.12.001
15. Wei W, Skye HM (2021) Residential net-zero energy buildings: review and perspective. Renew Sustain Energy Rev 142. https://doi.org/10.1016/j.rser.2021.110859
16. Bekele G, Tadesse G (2012) Feasibility study of small hydro/PV/wind hybrid system for off-grid rural electrification in Ethiopia. Appl Energy 97:5–15. https://doi.org/10.1016/j.apenergy.2011.11.059
17. Torcellini P, Grant R, Taylor C, Punjabi S, Diamond R, Colker R, Moy G, Kennett E. A common definition for zero energy buildings

18. Maduta C, Melica G, D'Agostino D, Bertoldi P (2022) Towards a decarbonised building stock by 2050: the meaning and the role of zero emission buildings (ZEBs) in Europe. Energy Strat Rev 44:101009. https://doi.org/10.1016/J.ESR.2022.101009
19. Sartori I, Napolitano A, Voss K (2012) Net zero energy buildings: a consistent definition framework. Energy Build 48:220–232. https://doi.org/10.1016/J.ENBUILD.2012.01.032
20. Galvin R (2022) Net-zero-energy buildings or zero-carbon energy systems? How best to decarbonize Germany's thermally inefficient 1950s–1970s-era apartments. J Build Eng 54. https://doi.org/10.1016/j.jobe.2022.104671
21. Abdelhafez MHH, Touahmia M, Noaime E, Albaqawy GA, Elkhayat K, Achour B, Boukendakdji M (2021) Integrating solar photovoltaics in residential buildings: towards zero energy buildings in Hail city, KSA. Sustainability (Switzerland) 13:1–19. https://doi.org/10.3390/SU13041845
22. López-Ochoa LM, Las-Heras-Casas J, González-Caballín JM, Carpio M (2023) Towards nearly zero-energy residential buildings in Mediterranean countries: the implementation of the energy performance of buildings directive 2018 in Spain. Energy 276. https://doi.org/10.1016/j.energy.2023.127539
23. Liu Z, Liu Y, He BJ, Xu W, Jin G, Zhang X (2019) Application and suitability analysis of the key technologies in nearly zero energy buildings in China. Renew Sustain Energy Rev 101:329–345. https://doi.org/10.1016/j.rser.2018.11.023
24. Lu Y, Wang S, Yan C, Shan K (2015) Impacts of renewable energy system design inputs on the performance robustness of net zero energy buildings. Energy 93:1595–1606. https://doi.org/10.1016/j.energy.2015.10.034
25. Woradechjumroen D, Yu Y, Li H, Yu D, Yang H (2014) Analysis of HVAC system oversizing in commercial buildings through field measurements. Energy Build 69:131–143. https://doi.org/10.1016/j.enbuild.2013.10.015
26. García-Sanz-Calcedo J, de Sousa Neves N, Almeida Fernandes JP (2021) Measurement of embodied carbon and energy of HVAC facilities in healthcare centers. J Clean Prod 289. https://doi.org/10.1016/j.jclepro.2020.125151
27. Mehrjerdi H, Hemmati R, Shafie-Khah M, Catalao JPS (2021) Zero energy building by multicarrier energy systems including hydro, wind, solar, and hydrogen. IEEE Trans Ind Inform 17:5474–5484. https://doi.org/10.1109/TII.2020.3034346
28. Heidari M, Rahdar MH, Dutta A, Nasiri F (2022) An energy retrofit roadmap to net-zero energy and carbon footprint for single-family houses in Canada. J Build Eng 60. https://doi.org/10.1016/j.jobe.2022.105141
29. Liu Z, Li M, Ji W (2025) Development and application of a digital twin model for net zero energy building operation and maintenance utilizing BIM-IoT integration. Energy Build 328:115170. https://doi.org/10.1016/J.ENBUILD.2024.115170
30. Abbasi S, Noorzai E (2021) The BIM-based multi-optimization approach in order to determine the trade-off between embodied and operation energy focused on renewable energy use. J Clean Prod 281. https://doi.org/10.1016/j.jclepro.2020.125359
31. Wu W, Skye HM (2018) Net-zero nation: HVAC and PV systems for residential net-zero energy buildings across the United States. Energy Convers Manag 177:605–628. https://doi.org/10.1016/j.enconman.2018.09.084
32. Kim D, Cho H, Koh J, Im P (2020) Net-zero energy building design and life-cycle cost analysis with air-source variable refrigerant flow and distributed photovoltaic systems. Renew Sustain Energy Rev 118. https://doi.org/10.1016/j.rser.2019.109508
33. Shaghaghi A, Eskandarpanah R, Gitifar S, Zahedi R, Pourrahmani H, Keshavarzzade M, Ahmadi A (2024) Energy consumption reduction in a building by free cooling using phase change material (PCM). Future Energy 3:31–36. https://doi.org/10.55670/FPLL.FUEN.3.2.4
34. Simson R, Thomsen KE, Wittchen KB, Kurnitski J (2021) A comparative analysis of NZEB energy performance requirements for residential buildings in Denmark, Estonia and Finland. E3S Web Conf 246. https://doi.org/10.1051/E3SCONF/202124614001
35. Abdous M, Aslani A, Noorollahi Y, Zahedi R, Yousefi H (2024) Design and analysis of zero-energy and carbon buildings with renewable energy supply and recycled materials. Energy Build 324:114922. https://doi.org/10.1016/J.ENBUILD.2024.114922

36. Giesekam J, Barrett JR, Taylor P (2016) Construction sector views on low carbon building materials. Build Res Inf 44:423–444. https://doi.org/10.1080/09613218.2016.1086872
37. Lu ML, Sun YJ, Kokogiannakis G, Ma ZJ (2024) Design of flexible energy systems for nearly/net zero energy buildings under uncertainty characteristics: a review. Renew Sustain Energy Rev 205:114828. https://doi.org/10.1016/J.RSER.2024.114828
38. Liu S, Cao S (2025) Development of integrated energy sharing systems between neighboring zero-energy buildings via micro-grid and local electric vehicles with energy trading business models. Appl Energy 380:124952. https://doi.org/10.1016/J.APENERGY.2024.124952
39. Chekired F, Taabli O, Khellili ZM, Tilmatine A, de Almeida AT, Canale L (2022) Near-zero-energy building management based on Arduino microcontroller—on-site lighting management application. Energies (Basel) 15. https://doi.org/10.3390/EN15239064
40. Kapsalaki M, Leal V (2011) Recent progress on net zero energy buildings. Adv Build Energy Res 5:129–162. https://doi.org/10.1080/17512549.2011.582352
41. Ionescu C, Baracu T, Vlad GE, Necula H, Badea A (2015) The historical evolution of the energy efficient buildings. Renew Sustain Energy Rev 49:243–253. https://doi.org/10.1016/J.RSER.2015.04.062
42. Romano G, Mancini F (2022) Transformation of a historical building into a nearly zero energy building (nZEB). J Phys Conf Ser 2385:012008. https://doi.org/10.1088/1742-6596/2385/1/012008
43. Ascanio-Villabona JG, Tarazona-Romero BE, Duran-Sarmiento MA, Lengerke O, Betancur-Arboleda LA (2023) Evaluation of the thermal performance of housing envelopes as passive cooling systems. Sustain Eng Innov 5:177–188. https://doi.org/10.37868/SEI.V5I2.ID227
44. Xiang X, Ma M, Ma X, Chen L, Cai W, Feng W, Ma Z (2022) Historical decarbonization of global commercial building operations in the 21st century. Appl Energy 322. https://doi.org/10.1016/j.apenergy.2022.119401
45. Al-Sukkari TAS, El-Daghar K, Galal KS, Afify A (2024) Towards effective implementation of building energy efficiency codes in Tripoli, Lebanon: key actions for enforcement. Herit Sustain Dev 6:771–790. https://doi.org/10.37868/HSD.V6I2.675
46. Esbensen TV, Korsgaard V (1977) Dimensioning of the solar heating system in the zero energy house in Denmark. Sol Energy 19:195–199. https://doi.org/10.1016/0038-092X(77)90058-5
47. Luo H, Cao S, Lu VL (2022) The techno-economic feasibility of a coastal zero-energy hotel building supported by the hybrid wind–wave energy system. Sustain Energy Grids Netw 30. https://doi.org/10.1016/j.segan.2022.100650
48. Attia S, Eleftheriou P, Xeni F, Morlot R, Ménézo C, Kostopoulos V, Betsi M, Kalaitzoglou I, Pagliano L, Cellura M, Almeida M, Ferreira M, Baracu T, Badescu V, Crutescu R, Hidalgo-Betanzos JM (2017) Overview and future challenges of nearly zero energy buildings (nZEB) design in Southern Europe. Energy Build 155:439–458. https://doi.org/10.1016/j.enbuild.2017.09.043
49. Durakovic B, Torlak M (2017) Experimental and numerical study of a PCM window model as a thermal energy storage unit. Int J Low-Carbon Technol 12:272–280. https://doi.org/10.1093/ijlct/ctw024
50. Fellah M, Ouhaibi S, Belouaggadia N, Mansouri K, Sebaibi N (2024) Enhancing building energy efficiency across diverse climatic zones in Morocco through optimized eco-friendly insulation thickness with air gap. Energy Rep 12:1551–1565. https://doi.org/10.1016/J.EGYR.2024.07.040
51. Sun Y (2024) The impact of green buildings on CO_2 emissions: evidence from commercial and residential buildings. J Clean Prod 469:143168. https://doi.org/10.1016/J.JCLEPRO.2024.143168
52. Dakwale VA, Ralegaonkar RV (2012) Review of carbon emission through buildings: threats, causes and solution. Int J Low-Carbon Technol 7:143–148. https://doi.org/10.1093/IJLCT/CTR032
53. Santamouris M, Vasilakopoulou K (2021) Present and future energy consumption of buildings: challenges and opportunities towards decarbonisation. e-Prime Adv Electr Eng Electron Energy 1:100002. https://doi.org/10.1016/J.PRIME.2021.100002

54. Proskuryakova L (2018) Updating energy security and environmental policy: energy security theories revisited. J Environ Manage 223:203–214. https://doi.org/10.1016/j.jenvman.2018.06.016
55. Wang Q, Zhou K (2017) A framework for evaluating global national energy security. Appl Energy 188:19–31. https://doi.org/10.1016/j.apenergy.2016.11.116
56. Calderon JL, Bazilian M, Sovacool B, Hund K, Jowitt SM, Nguyen TP, Månberger A, Kah M, Greene S, Galeazzi C, Awuah-Offei K, Moats M, Tilton J, Kukoda S (2020) Reviewing the material and metal security of low-carbon energy transitions. Renew Sustain Energy Rev 124. https://doi.org/10.1016/j.rser.2020.109789
57. Azab S, Rabie AE, Hafez F, Mostafa AH, El Rayes AH, Awad MM (2023) Decent life initiative and sustainable development goals: a systems thinking approach. Systems 11:446. https://doi.org/10.3390/SYSTEMS11090446
58. Günkaya Z, Özkan A, Banar M (2021) The effect of energy-saving options on environmental performance of a building: a combination of energy audit–life cycle assessment for a university building. Environ Sci Pollut Res 28:8822–8832. https://doi.org/10.1007/S11356-020-11141-Z
59. Li DHW, Yang L, Lam JC (2013) Zero energy buildings and sustainable development implications—a review. Energy 54:1–10. https://doi.org/10.1016/J.ENERGY.2013.01.070
60. Glennon J, Curran M, Spillane JP (2022) Nearly zero-energy buildings (nZEB) and their effect on social housing in Ireland: a case study review. In: Climate emergency—managing, building, and delivering the sustainable development goals, pp 59–69. https://doi.org/10.1007/978-3-030-79450-7_6
61. Tori F, Bustamante W, Vera S (2022) Analysis of net zero energy buildings public policies at the residential building sector: a comparison between Chile and selected countries. Energy Policy 161. https://doi.org/10.1016/j.enpol.2021.112707
62. Ozorhon B (2013) Response of construction clients to low-carbon building regulations. J Constr Eng Manag 139. https://doi.org/10.1061/(ASCE)CO.1943-7862.0000768
63. Zero-energy building (ZEB) market | global industry, 2032, https://www.businessresearchinsights.com/market-reports/zero-energy-building-zeb-market-102943. Accessed 11 Jan 2025
64. Wu Y, Zhong L (2023) An integrated energy analysis framework for evaluating the application of hydrogen-based energy storage systems in achieving net zero energy buildings and cities in Canada. Energy Convers Manag 286. https://doi.org/10.1016/j.enconman.2023.117066
65. Gaarder JE, Friis NK, Larsen IS, Time B, Møller EB, Kvande T (2023) Optimization of thermal insulation thickness pertaining to embodied and operational GHG emissions in cold climates—future and present cases. Build Environ 234. https://doi.org/10.1016/j.buildenv.2023.110187
66. Goldthau A (2011) Governing global energy: existing approaches and discourses. Curr Opin Environ Sustain 3:213–217. https://doi.org/10.1016/J.COSUST.2011.06.003
67. Gibbons L, Javed S (2022) A review of HVAC solution-sets and energy performance of nearly zero-energy multi-story apartment buildings in Nordic climates by statistical analysis of environmental performance certificates and literature review. Energy 238. https://doi.org/10.1016/j.energy.2021.121709
68. Nematchoua MK, Yvon A, Roy SEJ, Ralijaona CG, Mamiharijaona R, Razafinjaka JN, Tefy R (2019) A review on energy consumption in the residential and commercial buildings located in tropical regions of Indian Ocean: a case of Madagascar Island. J Energy Storage 24. https://doi.org/10.1016/j.est.2019.04.022
69. Avenali A, Catalano G, Giagnorio M, Matteucci G (2024) Factors influencing the adoption of zero-emission buses: a review-based framework. Renew Sustain Energy Rev 197:114388. https://doi.org/10.1016/J.RSER.2024.114388
70. Muratori M, Jadun P, Bush B, Bielen D, Vimmerstedt L, Gonder J, Gearhart C, Arent D (2020) Future integrated mobility-energy systems: a modeling perspective. Renew Sustain Energy Rev 119. https://doi.org/10.1016/J.RSER.2019.109541
71. Sustainability is more than a trend | zeb.nl. https://zeb-consulting.com/en-NL/sustainability-is-more-than-trend. Accessed 11 Jan 2025

Chapter 2
Building Codes and Energy Standards: Pathways Toward Zero Energy Buildings

Abbreviations

ASHRAE	American Society of Heating, Refrigerating and Air-Conditioning Engineers
BEC	Building Energy Certificate
DOE	Department of Energy
EPBD	European Building Code
EPC	Energy Performance Certificate
HVAC	Heating, Ventilation, and Air Conditioning
IDA ICE	Integrated Design and Assessment in Interactive Computer Environment
IEQ	Indoor Environmental Quality
LEED	Leadership in Energy and Environmental Design
NZEB	Net-zero Energy Building
nZEBs	Nearly Zero-energy Buildings
SEC	Smart Energy Communities
ZEB	Zero Energy Building

2.1 Introduction

Zero energy buildings (ZEBs) aim to minimize environmental impact by utilizing sustainably sourced materials and integrating renewable energy systems, striving for a balance between energy consumption and production [1]. Environmentally sustainable buildings, such as ZEBs, have proven to have multiple benefits for the environment, such as lowering energy use by close to 40% and lowering their carbon footprint and emissions by around 30% [2]. Governments all over the world have

B. Duraković, *Advancing Zero Energy Buildings*, Green Energy and Technology,
https://doi.org/10.1007/978-3-032-16601-2_2

begun to impose restrictions with the purpose of attaining sustainable development, such as regulating energy efficiency and establishing policies that have the potential to boost green building development [3]. What defines a net zero building is its energy efficiency that is achieved through the ability to accumulate energy from renewable sources, that are enough to sustain itself. Although legislation is important, the key factors are geographical location and the climatic conditions of the region. It is worth noting that ZEBs are not cost-efficient in the short term, but in the long term they are considered to have better economic performance [4].

Various topics of research are relevant when considering net zero, such as carbon emissions, decarbonization, type of building, ventilation, and other factors. The level of carbon emissions in the present is expected to have detrimental effects on the environment. As a result of that, the initiative for countries to implement green legislation has been improved. Consequently, countries have agreed to create zero emissions by the year 2050. Global targets can begin to be achieved by reducing carbon emissions and promoting the creation of solutions that will make it accessible for countries to implement net-zero strategies [5].

The definition of Net-Zero Energy Buildings (NZEBs) varies across the globe due to several factors, such as politics, economy, and others [6]. Therefore, there is currently no universally accepted standard for defining or characterizing a NZEB. Even the term itself is not consistently used, with various definitions and alternative labels in circulation. These include Net Zero Site Energy, Net Zero Source Energy, Net Zero Emissions, Net Zero Energy Costs, Net Zero Energy Commercial, and Nearly Net Zero Energy Building. Therefore, the specific interpretation adopted by a country often depends on its political goals, energy policies, and contextual conditions [4].

Some studies propose that the main focus of defining the concept of NZEBs should be buildings that are connected to an energy grid, rather than individual buildings with their autonomous source. Some experts argue that the term “Net” in NZEB should specifically denote buildings that are connected to the energy grid, distinguishing them from ZEBs, which may operate independently of the grid. The Net encompasses buildings that take and return energy from an energy grid over some time [7].

When considering developing and technologically advanced countries, energy consumption has increased due to various factors, such as growth in the economy and increase in world population. The spike in energy use has had many detrimental outcomes, such as the increase in the demand for energy, carbon emissions, acid rain, and global warming. Another resulting factor in the economic growth of countries is the need for new construction. New buildings are being made that are insufficiently sustainable but cannot be improved due to the prices of energy and insufficient innovation in technology [8]. Building energy codes can be used to improve the buildings and ensure that there are regulations that need to be fulfilled for both construction and design, which aim to be sustainable, conserve energy, and improve public health [9].

Building energy codes represent regulatory standards established by governmental authorities setting mandatory minimum requirements for the energy performance of new and renovated buildings, covering aspects such as insulation, Heating, Ventilation, and Air Conditioning (HVAC) systems, lighting, and overall energy use. Compliance with these codes is required for obtaining building permits and occupancy approvals.

Energy codes are a subset of building codes, which establish baseline requirements and govern building construction. Even though building standards have been more stringent in recent years, there has been an increase in noncompliance, particularly with energy-related building regulations [10]. Both in developing and developed countries, governments have gravitated towards the implementation of building codes to regulate the energy consumption of buildings and their performance. Those regulations have been considered key in maintaining good performance for the energy consumption of buildings [9, 11].

The purpose of this chapter is to provide an overview of existing studies on the building codes and standards in achieving net-zero energy targets. This chapter significantly advances the field of net zero energy by offering a robust framework of state-of-the-art tools and best practices that empower professionals and stakeholders to understand the concept of net zero energy in buildings. The goal is to illustrate the concept through presenting energy use through various existing building codes on international (or national) levels through the analysis of positive and negative sides and presenting advancements in the field.

2.2 Building Energy Code Overview

Building codes play a vital role in the construction and use of ZEBs, particularly in n regulating energy standards [12]. Building codes that integrate energy consumption regulation are used to define construction parameters, while also considering health, safety, and energy efficiency [13]. The building codes most often implemented in countries around the world are related to energy efficiency and often have action plans that use renewable energy, improve insulation efficiency, and use technologies that are energy efficient. Additionally, they regulate HVAC systems to make them energy efficient, use solar panels, utilize lighting that is energy efficient, and use sustainable composite materials during construction [14–16].

To achieve energy and climate objectives, countries adopt a national strategic document, e.g. The National Energy and Climate Plan (NECP) serve as a roadmap for transitioning towards a sustainable, low-carbon, and resilient energy system [17]. This document commonly sets minimum energy building efficiency requirements to support its broader climate objectives. For instance, in the UK, the Future Homes Standard, effective from 2025, the Net Zero Carbon Buildings Standard provides a framework to achieve net-zero carbon emissions in new and existing buildings and eliminating reliance on fossil fuels. Malaysia introduced the Energy Efficiency and Conservation Act (EECA) 2024 targeting energy intensive buildings to support Malaysia's goal of achieving carbon neutrality by 2050 [18–20].

Under the European Green Deal [21], the European Union aims to become the first climate-neutral continent by 2050, through the framework outlined in the Energy Performance of Buildings Directive (EPBD) [22]. Accordingly, all new buildings must be Nearly Zero-energy Buildings (nZEBs) and require member states to set long-term renovation strategies to decarbonize existing buildings. In support of this objective, financial support for research and innovation in building energy efficiency is provided through the Horizon Europe program [23] while the incentives for the adoption of energy-efficient technologies and practices are secured through the EU Emissions Trading System (ETS) [24].

Building energy codes in the United States are governed by the Department of Energy (DOE). Since 1991 Building Energy Codes Program supports the energy-efficient building codes to achieve cost-effective improvements in energy efficiency in new and renovated buildings through the International Energy Conservation Code and ASHRAE Standard 90.1 [20, 25]. The Building Energy Codes Program also administers programs such as the Building Codes Implementation for Efficiency and Resilience (BCIER) initiative, which provides and among the others supports like training, data collection and codes implementation [26]. The minimum energy efficiency requirements for new and existing buildings in Canada are governed by the National Energy Code of Canada for Buildings (NECB) [27]. The Pan-Canadian Framework on Clean Growth and Climate Change initiative includes measures to enhance building energy performance and support the development of a skilled workforce [28]. The minimum energy efficiency requirements for buildings in Australia are governed by the National Construction Code [29], released by the Australian Building Codes Board [30]. Australia has also committed to achieving net-zero emissions by 2050, through the initiatives such as the National Energy Productivity Plan [31], which aims to promote the adoption of energy-efficient technologies, and support the development of a skilled workforce in building energy efficiency.

The minimum energy performance standards for buildings in China are regulated by the Energy Conservation Law [32]. The goal is to achieve carbon neutrality by 2060, focusing on building sector as key areas through initiatives such as the Green Building Action Plan [33]. The initiative includes measures to promote the adoption of energy-efficient technologies, support the development of a skilled workforce, and promote building energy efficiency. The Energy Conservation Act is a minimum energy performance standard for buildings in Japan while the Top Runner Program is an initiative that targets for various energy-efficient products, technologies and practices, including building materials and equipment [34].

In Russia, the energy efficiency of buildings is primarily governed the Ministry of Construction, Housing and Utilities responsible for developing and implementing related policies, while the Federal Agency for Technical Regulation and Metrology is responsible for development of national standards related to building energy efficiency. The legal framework for energy efficiency in the Russia is regulated by Federal Law No. 261-FZ (2009) [35], while sets specific energy efficiency requirements for buildings is defined by the Ministerial Order No. 1550/pr (2017). For, GOST R

54,964–2012 is a standard that provides a conformity assessment method for environmental requirements in real estate, including energy efficiency aspects [36]. The Russian Energy Agency promotes energy efficiency and sustainable development in building sector [37].

Table 2.1 presents a comparative overview of the building sector energy efficiency and renewable energy targets across selected countries (regions), focusing on their goals for 2030 and 2050. This includes commitments related to new and existing buildings, renewable energy integration, and overall energy efficiency improvements.

Referring to Table 2.1, a global trend toward decarbonizing of the building sector is presented through the 2030 and 2050 goals for various countries and regions. Near-term (2030) goals are focused on improving energy efficiency through building codes, renovations, and increased renewable energy shares. Long-term (2050) goals are focused on net-zero or near-zero energy buildings and deep decarbonization of existing building stocks.

The EPBD sets out how European countries can achieve a decarbonized building stock by 2050. The EPBD aims for all new public buildings to meet this standard by 2027 and all new buildings by 2030. ZEBs are characterized by very low energy demand, fully met by renewable sources such as on-site generation, renewable energy communities, or district heating and cooling systems. In cases where these options are unavailable or limited, significant energy efficiency measures, both active and passive, may be necessary, starting with integrated solutions [68]. In Fig. 2.1 summarizes the key events and developments that will shape ZEB implementations.

The EPBD aims to accelerate large-scale renovation of the EU building stock, focusing on transitioning to emission-free heating systems, as currently only about 1% of buildings are renovated annually and 85–95% of existing buildings will still be standing in 2050. Despite progress over the past decade, inconsistent definitions and slow adoption of nZEB standards across Member States caused delays; thus, clear ZEB definitions are crucial to ensure timely implementation and contribute effectively to the EU's climate neutrality goals amid increasing urgency [69].

Most places aim to make new buildings near-zero energy or net-zero by 2030–2050 and to renovate existing buildings to meet higher energy standards. Renewable energy integration is a common priority, with many countries targeting significant increases in solar or clean energy use in buildings. While all aim for deep decarbonization by 2050, the level of ambition and specific approaches vary based on local conditions and development stages. Developed regions tend to set more aggressive renovation rates and strict efficiency mandates, whereas emerging economies focus more on gradual improvements and expanding renewable energy capacity. Overall, the goals reflect a shared global commitment to sustainable, low-carbon buildings to help meet climate targets.

Table 2.1 Comparative overview of ZEB goals by 2030 and 2050 per selected countries

Country	2030 goals	2050 goals	Ref.
EU	New buildings: All new buildings to be near-zero energy by 2030 Existing buildings: renovate the worst-performing 16% by 2030; 26% by 2033 Renewable Energy: Mandate solar systems for new buildings and major renovations between 2027 and 2030 Energy Efficiency: Achieve at least an 11.7% reduction in final energy consumption by 2030	New buildings: Achieve ZEB for all buildings by 2050 Existing buildings: implement 3% annual deep renovation rate for public buildings Renewable Energy: Mandate solar systems for new buildings and major renovations between 2027 and 2030 Energy Efficiency: Phase out fossil heating systems by 2040	[22, 38–43]
BiH	New Buildings: No mandatory ZEB obligations for new buildings by 2030 Existing Buildings: Retrofit existing stock at an annual rate of 0.8–1% Renewable Energy: Increase the share of renewable energy to approximately 43% of final energy consumption by 2030 Energy Efficiency: Reduce final energy consumption from 1,728,500 toe (2020) to 1,257,300 toe by 2030	New Buildings: No mandatory ZEB obligations for new buildings by 2050 Existing Buildings: Renovate up to 177,663 buildings by 2050 Renewable Energy: Increase the share of renewable energy to approximately 43% of final energy consumption by 2050 Energy Efficiency: Reduce final energy consumption from 1,728,500 toe (2020) to 1,257,300 toe by 2050	[44, 45]
China	New Buildings: All new buildings adhere to green building standards by 2025 Existing Buildings: Retrofit 350 million m^2 of existing buildings by 2025 Renewable Energy: Increase renewable energy share in building energy consumption to 8% by 2025 Energy Efficiency: Reduce energy intensity by 20% compared to 2015 levels	New Buildings: Achieve near-zero energy consumption for all buildings by 2050 Existing Buildings: Achieve renovation rates of 2.5% per year until 2030 and 3.5% until 2040 Renewable Energy: Increase renewable energy share in building energy consumption to 8% by 2025 Energy Efficiency: Reduce energy intensity by 45–50% compared to 2015 levels	[46, 47]

(continued)

Table 2.1 (continued)

Country	2030 goals	2050 goals	Ref.
Russia	New Buildings: Improve energy efficiency of buildings by 20% by 2030 compared to 2014 levels Existing Buildings: Implement energy efficiency measures in public sector organizations and apartment buildings Renewable Energy: Promote natural gas and nuclear power as key clean energy resources Energy Efficiency: Reduce energy intensity of GDP by 30% by 2030	New Buildings: Achieve a reduction of over three-quarters in direct CO_2 emissions from the building sector by 2050 Existing Buildings: Implement concrete and stringent building sector targets to reduce fossil fuel use and overall energy consumption Renewable Energy: Increase the share of non-fuel energy in the primary energy consumption from 11% up to 13–14% by 2030 Energy Efficiency: Achieve a reduction of over three-quarters in direct CO_2 emissions from the building sector by 2050	[37, 40, 48–51]
Japan	New Buildings: Mandate that all new houses and buildings comply with upgraded energy efficiency standards from 2025 Existing Buildings: Implement financial support for household renovations to improve energy efficiency Renewable Energy: Enhance the share of renewables to 50% of electricity needs by 2040 Energy Efficiency: Reduce greenhouse gas emissions by 60% by 2035 from 2013 levels	New Buildings: Achieve net-zero energy consumption for all buildings and houses by 2050 Existing Buildings: Implement financial support for household renovations to improve energy efficiency Renewable Energy: Enhance the share of renewables to 50% of electricity needs by 2040 Energy Efficiency: Achieve carbon neutrality by 2050	[52–54]
United Kingdom	*New Buildings:* All new homes to be zero carbon ready by 2025 Existing *Buildings:* Install 600,000 heat pumps per year by 2028 *Renewable Energy:* Increase the proportion of renewable energy in the national grid *Energy Efficiency*: Achieve EPC Band C for as many homes as possible by 2035	*New Buildings:* Achieve net-zero energy consumption for all buildings by 2050 *Existing Buildings*: Retrofit existing buildings to meet net-zero standards *Renewable Energy:* Achieve 100% clean electricity by 2035 *Energy Efficiency*: Achieve carbon neutrality by 2050	[55–57]

(continued)

Table 2.1 (continued)

Country	2030 goals	2050 goals	Ref.
Malaysia	*New Buildings*: No specific national targets for CO_2 reduction in buildings *Existing Buildings*: Implement energy efficiency measures in buildings Renewable Energy: Increase renewable energy capacity to 18,431 MW by 2040 Energy Efficiency: Achieve 10% energy efficiency savings in residential buildings by 2040	New Buildings: No specific national targets for CO_2 reduction in buildings Existing Buildings: Implement energy efficiency measures in buildings Renewable Energy: Achieve 70% renewable energy capacity by 2050 Energy Efficiency: Achieve 10% energy efficiency savings in residential buildings by 2040	[58, 59]
Canada lePara>	• Develop a "net-zero energy ready" model building code for adoption by provinces and territories • Retrofit existing buildings to meet higher energy performance standards • Achieve a 2% annual improvement in energy efficiency, doubling the global rate	• Achieve net-zero emissions in the building sector • Retrofit approximately 10 million buildings • Construct millions of new net-zero buildings	[60–62]
United States	• Federal blueprint aims for 65% reduction in building-sector GHG emissions by 2035, with interim goal of 50% emissions intensity cut by 2035 • All new federal buildings and major renovations to comply with zero emissions standards by 2030 • Federal Building Performance Standards require net zero Scope 1 emissions in 30% of federal buildings by 2030 • Building codes and tax credits via Inflation Reduction Act support wide-scale electrification, insulation, heatpump installations and grid integration	• Federal blueprint targets 90% emissions reduction in building sector by 2050 • Net zero emissions in all federal buildings by 2045 • DOE's national definition of "Zero Emissions Building" to standardize decarbonization efforts across public and private sectors	[63–65]
Brazil	Reduce direct CO_2 emissions from buildings by 33–48% by 2030. Increase the share of electricity in building energy consumption to 81%	Achieve net-zero emissions in the building sector by 2033–2034. Increase the share of electricity in building energy consumption to 95–98%	[66, 67]

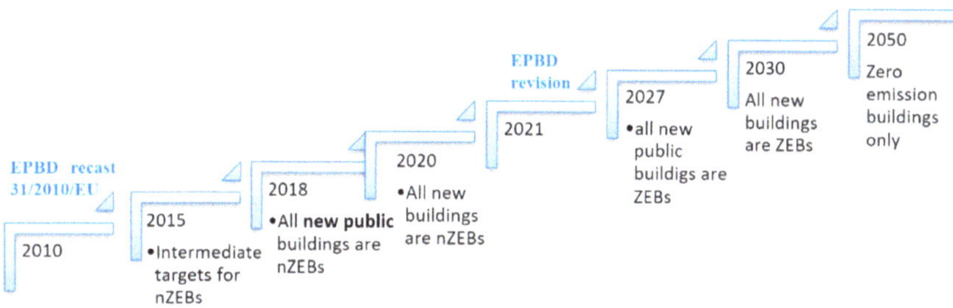

Fig. 2.1 Timeline for ZEB implementation

2.3 The Role of Energy Standards in Climate-Responsive Design

Local climatic conditions are critical for enhancing building energy efficiency since climate influences the energy needs that the facility generates. A developing pattern in the world goes towards an outlook that is net zero, which is achieved through integrating key strategies to achieve the desired results regarding energy use and demand [70]. Net zero energy buildings (zero energy building) share design aspects that encourage the goal of lowering or eliminating carbon emissions from the structure. Renewable energy is an important factor in achieving those results because it allows the building to produce its own energy, eliminating the need for the building to use alternative energy sources in the event of an increase in energy demand. After all, it is assumed that renewable energy sources will be stored according to calculations that define the capacity [71].

Green buildings are defined in the United States as structures that attempt to be efficient in all aspects, such as energy use, renewable and sustainable sourcing, and resource utilization over the building whole life cycle. This definition aligns with what green buildings mean globally, in countries such as China, South Africa, and Australia, which all have their building regulations and systems of certification. The implementation of such regulations prompts the developers, owners, and occupants of the buildings to raise their questions on how such building regulations influence the performance of the people using the building, the cost of the operation of the building, and how they return on the investment of developing them in the first place. As elements of design have an interactive quality, they influence the occupant of the building [72].

Countries such as China have been considered as ones with historically high pollution, and as such have harsh living conditions that may impact the public health of their residents. Consequently, China has implemented a variety of governance plans that induce action towards a cleaner environment. Those action plans show the government's willingness to combat pollution. While environmental regulations and action plans improve the quality of the environment. By assessing studies related to the subject, it was observed that environmental regulations do improve public health, mainly through the reduction of pollution and by implementing new innovative methods [73].

It is important to consider spaces that people occupy on a daily basis and how they affect energy demand. Indoor environment quality (IEQ) can be measured through various ways such as post-occupancy evaluations, or through a technique that uses systematic sampling, and compares an existing building with a conventional model that has the same specifications. Studies compare the models of conventional and green factories through various traits such as how clean it is, what furniture is available, the views, lighting, and others. The green factory is shown to be best in those attributes but lacks in the availability of ventilation, thermal comfort, control of the indoor environment, and others. Those drawbacks can be mitigated through implementing strategies that consider the climate of the location of the factory. There is a shortage of research on green buildings and how they work in the real world in various contexts and uses [74].

A multitude of factors influence tenant satisfaction in a facility such as an office building, including acoustic, thermal, air quality, workspace features, and others. The importance of occupant comfort is accounted to its relationship with overall productivity and performance at the workplace. Building sustainability can be enhanced by implementing sustainable measures that are governed by programs such as Leadership in Energy and Environmental Design (LEED), which is frequently used as a certification system for green buildings. While systems such as LEED are proven to be beneficial for the environment, there is lack of information about how they influence occupancy comfort. LEED is prominent in mainstream sustainability practices, and thus its relation to the comfort of the interior spaces of sustainable buildings for the occupants needs to be evaluated [75].

There are various legislations put in place by the EU regarding climate concerns and the building sector. The energy evolution of the sector was influenced by the implementation of zero-energy buildings [76]. Following that, the concept of zero-emission buildings was introduced, to ensure that there are no carbon emissions from residual fossil fuels. Positive energy can be achieved by different factors, such as strategies for battery cycling. Virtual models can be used to replicate different elements that contribute to building energy use, such as duration of occupancy, lighting, and the use of electric appliances [77].

A very important factor in lowering building energy consumption is to use sufficient insulation [78]. Additional strategies can be floor heating, which is more energy efficient than radiators and utilizes water that does not require heating for thermal comfort to be achieved. The geographic aspects of the location can be used as an advantage for net zero buildings, by using wind, solar, bioenergy, and other sources to generate more energy for the building to use [79]. The building sector attempts to fulfill carbon emission reduction targets are often separated into two paths: lowering energy usage through energy efficiency technologies and employing renewable energy sources to meet their own energy demands [80]. Some existing building codes encompass the design elements of buildings such as insulation.

2.4 Strengths and Weaknesses of Existing Regulations

Around 70% of carbon emissions are accounted for in urban areas in a country, and they are one of the most important targets in mitigating the impact of buildings on the environment. Cities are expected to become bigger with a rise in the number of the global population [81]. In developed economies, it is difficult to implement sustainable policies and reduce the environmental impact of construction because it is difficult to limit the amount of energy used in carbon emissions due to various constraints such as a lack of technical innovation, government regulations, geographical constraints, and others. Through prioritizing the most pressing concerns it can be made easier for contractors to implement sustainable elements [82].

When considering climate change, the building sector is responsible for around 37% of the global carbon emissions through research. Consequently, the logical next step is a partnership between stakeholders, that could provide a clear path leading to the achievement of net-zero buildings [83]. Stakeholders are considered an important factor due to their involvement in every step of the lifespan of a building. According to studies done on the topic, there are varying opinions on what signifies the start of a building's life, whether it is demolition or the building design process [84]. Reusing existing building sites and construction materials is considered a positive factor that contributes to sustainability [85]. It is critical to define the start of a building's lifespan to create precise regulations [86]. An important aspect in achieving near or net-zero buildings that have low emissions of greenhouse gases is beginning planning in the initial stages of the designing process of a building. The functioning and performance of a building should be designed and planned for it to have the least environmental impact it can have. To accurately analyze each stage of the design process, carbon emission calculations for a low-carbon building should be performed throughout the whole life cycle of the project [86]. Table 2.2 outlines some of the current effectiveness of building regulations in achieving net-zero emissions and identifies key gaps that hinder the progress.

Air quality and greenhouse gas emissions can be greatly lowered by regulating end-use equipment in various sectors such as residential, industrial, and commercial [94]. It is important to mention that it is difficult to make programs related to air quality energy efficient [86]. The regulatory frameworks related to the former are difficult to enforce due to the challenges faced when measuring or certifying the programs. As a result, there are only a few functional air quality systems that promote the fact they are energy efficient, thus they are not in widespread use [94]

Existing building standards are not community-based, they are more individual housing oriented and fail to provide ways to adopt new strategies for renewal energy addition that is community oriented. A recent case study in Mumbai, India, provides a detailed analysis of the existing Building Conservation Building Code compliance and a comparison with the newly proposed system [95]. The importance of this research for the theme is because it outlines ways that net-zero emissions could be achieved and provides a system that could be implemented into the building

Table 2.2 Effectiveness and challenges of existing regulations

Aspect	Current situation	Challenge	Source
Urban Emissions	Urban areas account for around 70% of global greenhouse gas emissions	Rapid urbanization increases emissions; existing regulations often lack scale and ambition	[81, 87]
Enforcement and Compliance	71 countries have mandatory energy codes; only 52 consistently enforce them	Enforcement is inconsistent, leading to only 30–40% of potential energy savings being realized	[88]
Retrofit Regulations	71 countries regulate new buildings; only 16 have comprehensive retrofit regulations	Lack of retrofit regulations despite 80% of future building stock already existing	[88]
Embodied Carbon Integration	Focus is mostly on operational energy use, but embodied carbon in building materials contribute to around 11% of all global carbon emissions	Embodied carbon (materials impact) often neglected missing holistic emissions accounting	[89]
Performance Gaps	Predicted vs actual energy performance often differ significantly	Some buildings use over 100% more energy than predicted due to occupant behavior and management	[90]
Financial Incentives and Awareness	High initial costs and limited awareness about NZEB benefits, especially in developing countries like South Africa	Financial and knowledge barriers limit adoption	[91]
Building Sector Emissions	The building sector is responsible for approximately 37% of global carbon emissions	Emission reductions are often limited by outdated codes and lack of integration across building lifecycles	[83, 92]
Stakeholder Involvement	Stakeholders are involved throughout a building's lifecycle	Unclear definitions of a building's lifespan complicate regulatory clarity and enforcement	[84, 93]

code which would be community-based and thus provides a more efficient solution, compared to the current individual-based strategies.

Office spaces that prioritize the well-being of their occupants and work towards inspiring productivity are key to acquiring IEQ. In some areas of the world, such as Taiwan, there are certification systems related to green buildings in place, that operate on a set of categories that serve as the criteria, such as sufficient ventilation, the acoustics of the building, energy-efficient lighting and decoration that impacts the occupant. One drawback of this system is that it does not cover the performance of thermal comfort in space. While most office buildings usually comply with the regulations set in place regarding acoustics or lighting, most of them exceed their pollution levels through excessive air conditioner use [96]. Green buildings tend to perform better according to the regulations compared to regular buildings in many areas such as the aforementioned, and additionally have better thermal comfort and quality of indoor air [97].

2.4.1 Global Perspectives and Challenges in Implementing Green Building Standards

Green buildings as a concept are considered to have existed from around the nineteenth century. As the idea of green buildings has differing definitions in the world, it is difficult to come to a unified set of regulations that would be enforced in the building sector worldwide [98]. Additionally, a lack of collaboration between researchers has emerged as a problem. Thus, it is difficult to come to a consensus regarding what universal law should be, as the capabilities of countries around the world differ greatly [99]. The concept of green buildings has existed, as previously mentioned, for decades before, but research journals have mostly emerged in the relatively recent past [100]. Building codes in countries throughout the world now mandate specific insulation standards for different building components, which are determined by their thermophysical qualities and calculated using established processes [101].

Building energy standards applied in several nations throughout the world aim to reduce energy use and carbon emissions. Global standards are aiming for net zero emissions. The European Building Code (EPBD) energy performance certificate highlights the need to accurately estimate thermal losses from the first design stages to achieve alignment between projected and real building energy usage. Nevertheless, several European country regulations and statutes regarding thermal insulation rely on rudimentary mathematical techniques that neglect or downplay the impact of thermal bridges, leading to significant disparities between projected and real heat losses across the building structure [101].

Some building codes, such as in Germany, regulate energy use at the source of the energy, to lower carbon emissions during operation. In countries such as Canada, regulations are focused on energy use and have regulations based on-site energy consumption [86]. Buildings of various types, such as residential, commercial, and office buildings, are consuming more energy and posing environmental concerns owing to their use of resources, how they affect the environment, and how this affects climate change [102]. Studies have shown that such buildings can account for up to 40% of the total energy use of countries. It is very important to enforce regulations regarding the energy efficiency of buildings at all functional levels, due to the increasing length of occupancy time of buildings, the growth of the population, and the increased need for new construction. The electric motors that power the systems use most of the energy utilized in buildings. In countries with increasing economic growth such as Malaysia, energy consumption has especially increased, particularly in residential and commercial buildings. In the fight against pollution, Malaysia has implemented various plans of action, such as regulations, constructing office buildings with low energy use, implementing renewable energy, and others [86].

Hong Kong can serve as an example for the implementation of building codes which limit energy consumption through setting levels of maximum energy consumption [103]. Those regulations encompass many different installations and the criterion for their implementation is based on performance. Studies have shown that strict implementations of building codes and regulations have a positive effect on the environment, reducing demand for energy, use of fuel, and an overall decrease in carbon emissions from energy use [104]. For building codes specifically, studies have shown that lowered greenhouse gas emissions accounted for the implementation of those very codes. To work towards conservation measures, additional building codes can be implemented, such as the Building Energy Code. The advantage of such a building code is that its criteria encompass the design of the building. Implementation could result in lower energy use and carbon emissions [105].

In some parts of the world such as Bahrain there can be an increase in the consumption of energy over a year. In the wider area around Bahrain electricity is the main source of energy. Studies show that buildings account for the most electricity use, over 50% [71]. The increase in electricity uses and demand is thought to be caused by an increase in the population. Other sectors such as commercial are expected to have an increase in electricity demand. Most of the electricity used comes from air-conditioning (AC) systems. By implementing energy codes, AC systems will have to be enhanced to be more efficient. Some regulations have been introduced such as window specifications, regarding its area and type of glazing, and additionally defining the maximum U-value of some elements of the buildings, such as the walls and roofs. Existing building codes have the limitation of not being applicable to a wide range of buildings, such as low-rise 1 or 2-story buildings [106]. Some countries, such as the UK, are legally bound to reduce carbon emissions by the year 2050. Around 30% of the energy demand is accounted for housing, and as such is a good candidate for carbon reduction. Achieving these steps by 2030, (especially decarbonizing new and existing buildings, implementing rigorous codes, electrifying systems, integrating renewables, and embracing lifecycle carbon thinking), is essential to reach mid-century net-zero and keep the 1.5 °C target within reach. Existing buildings should be refurbished to low carbon and architects should take the responsibility of designing sustainable buildings as a norm for all projects of all sizes, regardless of the type. Architects can face various challenges during the process. Still, sustainable construction should be planned from the design process, to ensure energy efficiency of the building [107].

2.5 Innovations and Compliance and Enforcement

Flexible environmental policies significantly enhance technological innovation, while environmental regulations positively moderate the link between flexible policies and sustainable development. However, there is a potential "implementation gap" between technical innovation and sustainable development [108]. Sustainable development has been best improved through innovation such as implementation of Phase Change Materials [109] and Smart Energy Communities (SEC). It is important to pay attention to what drives innovation and to nurture those aspects to promote it [110]. Clients have been considered one of the key drivers concerning the demand for sustainable development. Contrary to that, regulations have been the lowest in importance compared to other motivators. Through enhancing the process for clients to meet their demands, innovation can be promoted in construction firms. Innovation is complex due to the many factors that are related to each other during the process, such as demand for it, available materials and technology, government regulations and policies, and others [111].

The adoption of ZEB is accompanied by numerous complex challenges. Strategies must be developed to combat those challenges. The top priority when it comes to those challenges is the legislative barriers. Various issues can arise when trying to comply with legislation. Additionally, an important concern is the technical and professional limitations. A good approach to making the process seamless is to promote collaboration between all the parties involved including government, the building sector, the owners of the buildings, and the others [112].

SEC has emerged as a response to the growing emphasis on energy, environmental, and social sustainability. These communities are designed to integrate intelligent technologies and practices for local energy development. From a linguistic perspective, the term "community" denotes a social entity characterized by shared practices, values, and a sense of belonging [113]. Some innovative methods include artificial intelligence, genetic algorithms, and Integrated Design and Assessment in Interactive Computer Environment (IDA ICE). Artificial intelligence can be used to simulate models of energy-efficient units based on various techniques, such as using a binary gravitational search algorithm. Various renewable energy sources, such as solar, battery, and other systems, can be employed and integrated to reduce or eliminate energy production and consumption expenses. The use of such measures can transform buildings into net zero buildings. Battery banks can be used to manage the power that is in use in the building, while capacitor banks can be used to manage the power needed for a reaction that is in the system. To lower the energy costs of the net zero building, it must produce more energy than it consumes [114, 115]. Genetic algorithms and IDA ICE can be used to create sustainable and low-cost buildings. Table 2.3 summarizes key innovation types used in sustainable architecture, their specific applications, intended goals, and supporting references.

Referring to Table 2.3 it can be observed that innovative technologies are increasingly being applied in the field of sustainable building design to meet environmental and energy efficiency goals. Technologies such as Artificial Intelligence, Genetic

Table 2.3 Innovative technologies for sustainable building design

Innovation	Application	Goals	Ref.
Artificial intelligence	Simulating models of energy-efficient units	Transforming buildings into net-zero	[116]
Genetic algorithms	Sustainable, low-cost buildings; combined with IDA ICE	Lowering carbon emissions and costs by about 32%	[117]
IDA ICE	Various sectors to model and optimize building performance, focusing on energy efficiency, indoor climate, and environmental sustainability	Sustainability, reducing energy demand	[118]

Algorithms, and simulation tools like IDA ICE offer advanced methods for analyzing and optimizing building performance. These innovative technologies support the shift toward energy efficient solutions addressing some of the sustainability challenges including energy demand reduction and lowering operational costs. This reflects a broader trend in the construction industry, where digital tools are playing a crucial role in driving environmental responsibility and long-term efficiency.

2.6 Building Energy Certificate

Building Energy Certificates (BECs), such as Energy Performance Certificate (EPC) is a comprehensive assessment tool that evaluates a building's energy efficiency and environmental impact. It includes an Energy Efficiency Index, typically rated from A (most efficient) to G (least efficient) [119], which reflects the building energy consumption relative to building envelope insulation, heating/cooling systems, and electrical appliances. Additionally, EPCs provide a detailed analysis of energy use across various systems, such as heating, cooling, and lighting, enabling property owners to identify areas for improvement. The document precisely states how much energy a building consumes, how much energy it unnecessarily loses, and which energy class it belongs to. They also offer indicators for carbon emissions, highlighting the environmental footprint of the building. The certificate also contains a proposal for measures to take to move a building into a higher energy class, with an estimate of energy savings and the time frame in which these measures will pay off.

The purpose of an energy certificate is to provide information to the owner, future buyer or user of the space about how much energy a building consumes and how much it should consume. The energy class designation is mandatory in the advertisement for the sale of real estate, and it is also part of the purchase contract.

Some BECs integrate smart technologies, allowing real-time monitoring of energy performance through systems like Building Information Modelling and Geographic Information Systems [120]. Furthermore, BECs present visual results, often using color-coded bands, to facilitate quick understanding and comparison of energy

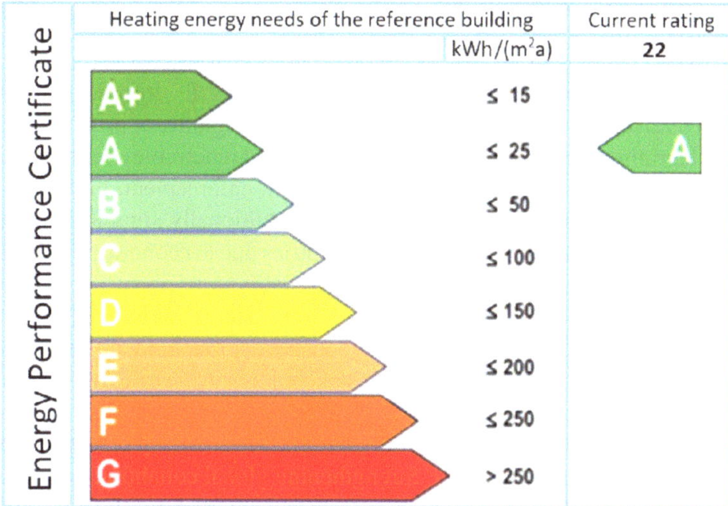

Fig. 2.2 Example of energy performance certificate

performance [121, 122]. In many regions, obtaining an EPC is a legal requirement when selling, renting, or constructing a building, ensuring transparency and compliance with energy efficiency standards [123, 124]. Figure 2.2 provides an example of a building energy certificate illustrating the building energy performance with a color-coded scale from A + to G.

Energy certificates are issued by authorized bodies / people registered for energy audits and energy certification of buildings. Buildings are classified into eight energy classes (from A + to G), with the highest class (A +) indicating buildings that consume less than 15 kWh/m2, and higher classes indicating buildings that consume more than 250 kW/m^2. Class A + includes the most energy-efficient buildings, such as passive houses or ZEBs. These buildings are characterized by excellent insulation properties, the use of heat recovery systems and renewable energy sources. For instance, class A and B may include low-energy buildings that have very energy efficient envelopes and use renewable energy sources etc., while classes F and G may include older poorly insulated buildings. Therefore, to achieve a higher class, envelope insulation and the use of renewable energy sources are necessary.

The benefits of Building Energy Certificates extend beyond regulatory compliance. Properties with higher EPC ratings are often more attractive to potential buyers and tenants, as they typically promise lower energy bills and a reduced environmental impact. This can lead to increased market value and rental yields. Moreover, EPCs can guide property owners in making informed decisions about energy-saving measures, leading to long-term cost savings. They also play a crucial role in achieving sustainability goals, contributing to a property's overall environmental responsibility and helping to mitigate climate change [125, 126].

2.7 Conclusion

Building energy codes and standards are pivotal tools in driving the transition towards net-zero energy buildings worldwide. While significant progress has been made in integrating energy efficiency requirements and renewable energy provisions, numerous challenges remain, including inconsistent enforcement, limited focus on embodied carbon, and gaps in retrofit regulations. Additionally, significant challenges in achieving these goals exist in developing countries due to economic instability and rapid population growth.

Therefore, to fully realize the potential of zero energy buildings, future codes must adopt a more holistic and climate-responsive approach, incorporate lifecycle assessments, and engage a broader range of stakeholders. All parties involved such as governments, contractors, investors, and others must come together to work towards improving current energy use standards and aim towards implementing net-zero strategies for a cleaner environment. Strengthening global collaboration and knowledge sharing will be essential to harmonize standards and accelerate sustainable building practices. Ultimately, evolving building codes into dynamic, adaptive frameworks will enable resilient, low-carbon built environments that contribute effectively to global climate goals.

References

1. Garde F, Lenoir A, Scognamiglio A, Aelenei D, Waldren D, Rostvik HN, Ayoub J, Aelenei L, Donn M, Tardif M, Cory S (2014) Design of net zero energy buildings: feedback from international projects. Energy Procedia 61:995–998. https://doi.org/10.1016/J.EGYPRO.2014.11.1011
2. Jaysawal RK, Chakraborty S, Elangovan D, Padmanaban S (2022) Concept of net zero energy buildings (NZEB)—A literature review. Clean Eng Technol 11:100582. https://doi.org/10.1016/J.CLET.2022.100582
3. He W, Zhang Y, Kong D, Li S, Wu Z, Zhang L, Liu P (2024) Promoting green-building development in sustainable development strategy: a multi-player quantum game approach. Expert Syst Appl 240:122218. https://doi.org/10.1016/J.ESWA.2023.122218
4. Cielo D, Subiantoro A (2021) Net zero energy buildings in New Zealand: challenges and potentials reviewed against legislative, climatic, technological, and economic factors. J Buil Eng 44:102970. https://doi.org/10.1016/J.JOBE.2021.102970
5. Navarro Bringas E, Godawatte GAGR (2022) Shedding light on the efforts into the rehabilitation of a major culprit of carbon emissions: a scientometric analysis of net-zero in the built environment sector. Energy Build 266:112119. https://doi.org/10.1016/J.ENBUILD.2022.112119
6. Harkouss F, Fardoun F, Biwole PH (2019) Optimal design of renewable energy solution sets for net zero energy buildings. Energy 179:1155–1175. https://doi.org/10.1016/J.ENERGY.2019.05.013
7. Sartori I, Napolitano A, Voss K (2012) Net zero energy buildings: a consistent definition framework. Energy Build 48:220–232. https://doi.org/10.1016/J.ENBUILD.2012.01.032
8. Fanney AH, Payne V, Ullah T, Ng L, Boyd M, Omar F, Davis M, Skye H, Dougherty B, Polidoro B, Healy W, Kneifel J, Pettit B (2015) Net-zero and beyond! Design and performance

of NIST's net-zero energy residential test facility. Energy Build 101:95–109. https://doi.org/10.1016/J.ENBUILD.2015.05.002
9. Iwaro J, Mwasha A (2010) A review of building energy regulation and policy for energy conservation in developing countries. Energy Policy 38:7744–7755. https://doi.org/10.1016/J.ENPOL.2010.08.027
10. Fujimoto T, Yamaguchi Y, Shimoda Y (2017) Energy management for voltage control in a net-zero energy house community considering appliance operation constraints and variety of households. Energy Build 147:188–199. https://doi.org/10.1016/J.ENBUILD.2017.05.009
11. Al-Sukkari TAS, El-Daghar K, Galal KS, Afify A (2024) Towards effective implementation of building energy efficiency codes in Tripoli, Lebanon: key actions for enforcement. Herit Sustain Dev 6:771–790. https://doi.org/10.37868/HSD.V6I2.675
12. Foroushani S, Bernhardt R, Bernhardt M (2022) On the use of the reference building approach in modern building energy codes. Energy Build 256:111726. https://doi.org/10.1016/J.ENBUILD.2021.111726
13. Omrany H, Chang R, Soebarto V, Zhang Y, Ghaffarianhoseini A, Zuo J (2022) A bibliometric review of net zero energy building research 1995–2022. Energy Build 262:111996. https://doi.org/10.1016/J.ENBUILD.2022.111996
14. Janda KB, Busch JF (1994) Worldwide status of energy standards for buildings. Energy 19:27–44. https://doi.org/10.1016/0360-5442(94)90102-3
15. Findik F (2025) Sustainable composite materials: applications and future. Sustain Eng Innov 7:415–424. https://doi.org/10.37868/SEI.V7I2.ID504
16. Durakovic B, Mujković N, Hudović-Kljuno A, Gibanica E (2025) Effect of phase change materials on the thermal behavior of biomass-based composites. Herit Sustain Dev 7:155–166. https://doi.org/10.37868/HSD.V7I1.1107
17. In focus: national energy and climate plans–European commission. https://commission.europa.eu/news-and-media/news/focus-national-energy-and-climate-plans-2020-06-16_en. Accessed 24 June 2025
18. Malaysia to implement EECA in 2025, targeting energy-intensive buildings|NEWS|Reccessary. https://www.reccessary.com/en/news/malaysia-s-eeca-to-boost-decarbonization-of-buildings. Accessed 24 June 2025
19. Malaysia publishes energy efficiency and conservation act|Enviliance ASIA. https://enviliance.com/regions/southeast-asia/my/report_12837?utm_source=chatgpt.com. Accessed 24 June 2025
20. Codes 101|Building energy codes program. https://energycodes.gov/codes-101?utm_source=chatgpt.com. Accessed 24 June 2025
21. Energy efficiency targets. https://energy.ec.europa.eu/topics/energy-efficiency/energy-efficiency-targets-directive-and-rules/energy-efficiency-targets_en?utm_source=chatgpt.com. Accessed 24 June 2025
22. Energy performance of buildings directive. https://energy.ec.europa.eu/topics/energy-efficiency/energy-efficient-buildings/energy-performance-buildings-directive_en?utm_source=chatgpt.com. Accessed 24 June 2025
23. Current funding. https://energy.ec.europa.eu/topics/energy-efficiency/financing/current-funding_en. Accessed 24 June 2025
24. EU emissions trading system (EU ETS)–European commission. https://climate.ec.europa.eu/eu-action/eu-emissions-trading-system-eu-ets_en?utm_source=chatgpt.com. Accessed 24 June 2025
25. Technical assistance network|Building energy codes program. https://energycodes.gov/technical-assistance/network?utm_source=chatgpt.com. Accessed 24 June 2025
26. Building codes implementation for efficiency and resilience|Department of energy. https://www.energy.gov/eere/building-codes-implementation-efficiency-and-resilience?utm_source=chatgpt.com. Accessed 24 June 2025
27. National energy code of Canada for buildings 2020–National research council Canada. https://nrc.canada.ca/en/certifications-evaluations-standards/codes-canada/codes-canada-publications/national-energy-code-canada-buildings-2020. Accessed 24 June 2025

28. Pan-Canadian framework on clean growth and climate change: strategic environmental assessment–Canada.ca. https://www.canada.ca/en/environment-climate-change/corporate/transparency/strategic-environmental-economic-assessments/pan-canadian-framework.html. Accessed 24 June 2025
29. NCC 2022 adopted|ABCB. https://www.abcb.gov.au/news/2023/ncc-2022-adopted?utm_source=chatgpt.com. Accessed 24 June 2025
30. Australian building codes board|ABCB. https://www.abcb.gov.au/. Accessed 24 June 2025
31. National energy productivity plan–DCCEEW. https://www.dcceew.gov.au/energy/publications/national-energy-productivity-plan?utm_source=chatgpt.com. Accessed 24 June 2025
32. Energy conservation law of the People's Republic of China (Amended in 2018). https://www.lexiscn.com/law/law-english-1-3343319-T.html?utm_source=chatgpt.com. Accessed 24 June 2025
33. Energy-efficient buildings in China: Standards and financing mechanisms–Sino-German cooperation on climate change. https://climatecooperation.cn/climate/energy-efficient-buildings-in-china-standards-and-financing-mechanisms/?utm_source=chatgpt.com. Accessed 24 June 2025
34. The government enacted the revised Building Energy Conservation Act–ECCJ/Asia energy efficiency and conservation collaboration center. https://www.asiaeec-col.eccj.or.jp/policynews-202207/?utm_source=chatgpt.com. Accessed 24 June 2025
35. Federal law No. 261-FZ "on energy conservation and increase of energy efficiency"—policies–IEA. https://www.iea.org/policies/1749-federal-law-no-261-fz-on-energy-conservation-and-increase-of-energy-efficiency. Accessed 24 June 2025
36. Grafkina M, Sviridova E (2016) New approaches to the electromagnetic monitoring for improving the regulatory framework "green" building. MATEC Web of Conferences 86. https://doi.org/10.1051/MATECCONF/20168607004
37. Alekseev AN, Lobova SV, Bogoviz AV, Ragulina YV (2019) A critical review of Russia's energy efficiency policies in the construction and housing sector. Int J Energy Econ Policy 9:166–172
38. Nearly-zero energy and zero-emission buildings–European Commission. https://energy.ec.europa.eu/topics/energy-efficiency/energy-efficient-buildings/nearly-zero-energy-and-zero-emission-buildings_en?utm_source=chatgpt.com. Accessed 24 June 2025
39. Energy performance of buildings directive EU/2024/1275. https://eur-lex.europa.eu/legal-content/EN/TXT/HTML/?uri=OJ:L_202401275
40. Public buildings. https://energy.ec.europa.eu/topics/energy-efficiency/energy-efficiency-targets-directive-and-rules/public-buildings_en?utm_source=chatgpt.com. Accessed 24 June 2025
41. Energy performance of buildings–bpv BRAUN PARTNERS. https://www.bpv-bp.com/en/clanek/energy-performance-of-buildings/?utm_source=chatgpt.com. Accessed 24 June 2025
42. Babiarz B, Krawczyk DA, Siuta-Olcha A, Manuel CD, Jaworski A, Barnat E, Cholewa T, Sadowska B, Bocian M, Gnieciak M, Werner-Juszczuk A, Kłopotowski M, Gawryluk D, Stachniewicz R, Święcicki A, Rynkowski P (2024) Energy efficiency in buildings: toward climate neutrality. Energies 17:4680. https://doi.org/10.3390/EN17184680
43. Towards zero-emission buildings by 2050: council adopts rules to improve energy performance–Consilium. https://www.consilium.europa.eu/en/press/press-releases/2024/04/12/towards-zero-emission-buildings-by-2050-council-adopts-rules-to-improve-energy-performance/?utm_source=chatgpt.com. Accessed 24 June 2025
44. Bosnia and Herzegovina's renovation strategy for 2050|BUILD UP. https://build-up.ec.europa.eu/en/news-and-events/news/bosnia-and-herzegovinas-renovation-strategy-2050. Accessed 24 June 2025
45. Bosnia and Herzegovina. https://1p5ndc-pathways.climateanalytics.org/countries/bosnia-and-herzegovina/. Accessed 24 June 2025
46. CO_2 and energy targets|China|Global sustainable buildings guide|Baker McKenzie resource hub. https://resourcehub.bakermckenzie.com/en/resources/global-sustainable-buildings-index/asia-pacific/china/topics/co2-and-energy-targets. Accessed 24 June 2025

47. Policies & action|Climate action tracker. https://climateactiontracker.org/countries/china/2023-11-22/policies-action/. Accessed 24 June 2025
48. Energy strategy of Russia to 2030—Policies–IEA. https://www.iea.org/policies/1370-energy-strategy-of-russia-to-2030?utm_source=chatgpt.com. Accessed 24 June 2025
49. Policies & action|Climate action tracker. https://climateactiontracker.org/countries/russian-federation/policies-action/?utm_source=chatgpt.com. Accessed 24 June 2025
50. Steblyakova LP, Vechkinzova E, Khussainova Z, Zhartay Z, Gordeyeva Y (2022) Green energy: new opportunities or challenges to energy security for the common electricity market of the Eurasian economic union countries. Energies 15:5091. https://doi.org/10.3390/EN15145091
51. Buildings. https://1p5ndc-pathways.climateanalytics.org/countries/russia/sectors/buildings?utm_source=chatgpt.com. Accessed 24 June 2025
52. Japan aims to cut greenhouse gas emissions 60% by 2035 vs 2013 levels|Reuters. https://www.reuters.com/world/japan/japan-aims-cut-greenhouse-gas-emissions-60-by-2035-vs-2013-levels-2024-12-24/?utm_source=chatgpt.com. Accessed 24 June 2025
53. Japan targets 40–50% power supply from renewables by 2040|Reuters. https://www.reuters.com/sustainability/climate-energy/japan-targets-40-50-power-supply-renewable-energy-by-2040-2024-12-17/?utm_source=chatgpt.com. Accessed 24 June 2025
54. MLIT expands support for energy-efficient buildings/Revised law to take effect–ECCJ/Asia energy efficiency and conservation collaboration center. https://www.asiaeec-col.eccj.or.jp/policynews-202408-1/?utm_source=chatgpt.com. Accessed 24 June 2025
55. Home|UK net zero carbon buildings standard. https://www.nzcbuildings.co.uk/. Accessed 24 June 2025
56. Niel J (2021) Net zero strategy: build back greener. Gov.Uk. 368
57. Slater T, Hogg AE, Mottram R (2020) Ice-sheet losses track high-end sea-level rise projections. Nat Clim Chang 10:879–881. https://doi.org/10.1038/S41558-020-0893-Y
58. Malaysia's green business building opportunity|McKinsey
59. Malaysia's commitment to achieving net-zero emissions by 2050—SDG Malaysia
60. Government of Canada's new Green Buildings Strategy: a plan to help Canadians save money on their energy bills–Canada.ca. https://www.canada.ca/en/department-justice/news/2024/07/government-of-canadas-new-green-buildings-strategy-a-plan-to-help-canadians-save-money-on-their-energy-bills.html?utm_source=chatgpt.com. Accessed 24 June 2025
61. Canada's 2030 emissions reduction plan–Chapter 2–Canada.ca. https://www.canada.ca/en/services/environment/weather/climatechange/climate-plan/climate-plan-overview/emissions-reduction-2030/plan/chapter-2.html?utm_source=chatgpt.com. Accessed 24 June 2025
62. Net-zero energy ready buildings in Canada–Efficiency Canada. https://www.efficiencycanada.org/building-codes/net-zero-energy-ready-buildings-in-canada/?utm_source=chatgpt.com. Accessed 24 June 2025
63. Net-zero emissions buildings by 2045, including a 50% reduction by 2032|Federal sustainability plan|Office of the federal chief sustainability officer. https://www.sustainability.gov/archive/biden46/federalsustainabilityplan/buildings.html?utm_source=chatgpt.com. Accessed 24 June 2025
64. DOE releases blueprint to decarbonize America's buildings sector. https://www.mcmorrowreports.com/doe-releases-first-ever-federal-blueprint-to-decarbonize-americas-buildings-sector/?utm_source=chatgpt.com. Accessed 24 June 2025
65. Federal building performance standard|Office of the federal chief sustainability officer. https://www.sustainability.gov/archive/biden46/federalbuildingstandard.html?utm_source=chatgpt.com. Accessed 24 June 2025
66. Climate analytics|Brazil not on track to achieve 2030 target or…. https://climateanalytics.org/publications/15c-aligned-2030-and-2035-ghg-emissions-targets/brazil?utm_source=chatgpt.com. Accessed 24 June 2025
67. Bezerra P, da Silva F, Cruz T, Mistry M, Vasquez-Arroyo E, Magalar L, De Cian E, Lucena AFP, Schaeffer R (2021) Impacts of a warmer world on space cooling demand in Brazilian households. Energy Build 234. https://doi.org/10.1016/J.ENBUILD.2020.110696

68. Fotiou T, Fragkos P, Zisarou E (2024) Decarbonising the EU buildings|Model-based insights from European countries. Climate 12:85. https://doi.org/10.3390/CLI12060085
69. Maduta C, Melica G, D'Agostino D, Bertoldi P (2022) Towards a decarbonised building stock by 2050: the meaning and the role of zero emission buildings (ZEBs) in Europe. Energ Strat Rev 44:101009. https://doi.org/10.1016/J.ESR.2022.101009
70. D'Agostino D, Parker D (2018) A framework for the cost-optimal design of nearly zero energy buildings (NZEBs) in representative climates across Europe. Energy 149:814–829. https://doi.org/10.1016/J.ENERGY.2018.02.020
71. Feng W, Zhang Q, Ji H, Wang R, Zhou N, Ye Q, Hao B, Li Y, Luo D, Lau SSY (2019) A review of net zero energy buildings in hot and humid climates: experience learned from 34 case study buildings. Renew Sustain Energy Rev 114:109303. https://doi.org/10.1016/J.RSER.2019.109303
72. Thatcher A, Milner K (2016) Is a green building really better for building occupants? A longitudinal evaluation. Build Environ 108:194–206. https://doi.org/10.1016/J.BUILDENV.2016.08.036
73. Rauf A, Zhong K, Zhao X, Hu H, Pei Y, Chen H, Liu Z. Does environmental regulation improve residents' health? Evidence from China
74. Ravindu S, Rameezdeen R, Zuo J, Zhou Z, Chandratilake R (2015) Indoor environment quality of green buildings: case study of an LEED platinum certified factory in a warm humid tropical climate. Build Environ 84:105–113. https://doi.org/10.1016/J.BUILDENV.2014.11.001
75. Altomonte S, Schiavon S (2013) Occupant satisfaction in LEED and non-LEED certified buildings. Build Environ 68:66–76. https://doi.org/10.1016/J.BUILDENV.2013.06.008
76. Boitier B, Nikas A, Gambhir A, Koasidis K, Elia A, Al-Dabbas K, Alibaş Ş, Campagnolo L, Chiodi A, Delpiazzo E, Doukas H, Fougeyrollas A, Gargiulo M, Le Mouël P, Neuner F, Perdana S, van de Ven D-J, Vielle M, Zagamé P, Mittal S (2023) A multi-model analysis of the EU's path to net zero. Joule. https://doi.org/10.1016/J.JOULE.2023.11.002
77. Bilardo M, Kämpf JH, Fabrizio E (2024) From zero energy to zero power buildings: a new paradigm for a sustainable transition of the building stock. Sustain Cities Soc 101:105136. https://doi.org/10.1016/J.SCS.2023.105136
78. Jabari F, Ghaebi H, Mohammadi-Ivatloo B, Mohammadpourfard M, Bannae-Sharifian MB (2020) Design, worst case study, and sensitivity analysis of a net-zero energy building for sustainable urban development. Sustain Cities Soc 54:101991. https://doi.org/10.1016/J.SCS.2019.101991
79. Bistline JET, Blanford GJ (2021) The role of the power sector in net-zero energy systems. Energy Clim Chang 2:100045. https://doi.org/10.1016/J.EGYCC.2021.100045
80. Ahmed A, Ge T, Peng J, Yan WC, Tee BT, You S (2022) Assessment of the renewable energy generation towards net-zero energy buildings: a review. Energy Build 256:111755. https://doi.org/10.1016/J.ENBUILD.2021.111755
81. Kilkis B (2022) Net-zero buildings, what are they and what they should be? Energy 256:124442. https://doi.org/10.1016/J.ENERGY.2022.124442
82. Madhusanka HWN, Pan W, Kumaraswamy MM (2022) Constraints to low-carbon building: perspectives from high-rise high-density cities. Energy Build. 275:112497. https://doi.org/10.1016/J.ENBUILD.2022.112497
83. Vine EL (1996) Residential building code compliance: Implications for evaluating the performance of utility residential new construction programs. Energy 21:1051–1058. https://doi.org/10.1016/0360-5442(96)00076-X
84. Falana J, Osei-Kyei R, Tam VW (2024) Towards achieving a net zero carbon building: a review of key stakeholders and their roles in net zero carbon building whole life cycle. J Build Eng 82:108223. https://doi.org/10.1016/J.JOBE.2023.108223
85. Aksamija A (2016) Regenerative design and adaptive reuse of existing commercial buildings for net-zero energy use. Sustain Cities Soc 27:185–195. https://doi.org/10.1016/J.SCS.2016.06.026
86. Karlsson I, Rootzén J, Johnsson F, Erlandsson M (2021) Achieving net-zero carbon emissions in construction supply chains—A multidimensional analysis of residential building systems. Dev Built Environ 8:100059. https://doi.org/10.1016/J.DIBE.2021.100059

87. Cities and climate change|UNEP–UN environment programme. https://www.unep.org/ar/node/1411. Accessed 20 June 2025
88. Moving the needle on building energy codes: from analysis to action. https://blogs.worldbank.org/en/developmenttalk/moving-the-needle-on-building-energy-codes--from-analysis-to-act. Accessed 20 June 2025
89. The problem with net zero: embodied carbon–world green building council. https://worldgbc.org/climate-action/trends-and-innovations/embodied-carbon/. Accessed 20 June 2025
90. Shi X, Si B, Zhao J, Tian Z, Wang C, Jin X, Zhou X (2019) Magnitude, causes, and solutions of the performance gap of buildings: a review. Sustainability 11:937. https://doi.org/10.3390/SU11030937
91. Terblanche R, May C, Steward J (2025) Implementing and operating net zero buildings in South Africa. Build Cities 6:255–271. https://doi.org/10.5334/BC.549
92. The building sector is key to the fight against climate change|World economic forum. https://www.weforum.org/stories/2024/06/building-sector-climate-change-construction-materials/. Accessed 20 June 2025
93. Advancing climate action through an Urban Lens. https://www.worldbank.org/en/topic/urbandevelopment/brief/climate-action-through-an-urban-lens. Accessed 20 June 2025
94. Vine E (2003) Opportunitites for promoting energy efficiency in buildings as an air quality compliance approach. Energy 28:319–341. https://doi.org/10.1016/S0360-5442(02)00112-3
95. Rethnam OR, Thomas A (2023) A community building energy modelling—Life cycle cost analysis framework to design and operate net zero energy communities. Sustain Prod Consum 39:382–398. https://doi.org/10.1016/J.SPC.2023.04.022
96. Pang Z, O'Neill Z, Chen Y, Zhang J, Cheng H, Dong B (2023) Adopting occupancy-based HVAC controls in commercial building energy codes: analysis of cost-effectiveness and decarbonization potential. Appl Energy 349:121594. https://doi.org/10.1016/J.APENERGY.2023.121594
97. Liang HH, Chen CP, Hwang RL, Shih WM, Lo SC, Liao HY (2014) Satisfaction of occupants toward indoor environment quality of certified green office buildings in Taiwan. Build Environ 72:232–242. https://doi.org/10.1016/J.BUILDENV.2013.11.007
98. Lin B, Chen Z (2022) Net zero energy building evaluation, validation and reflection—A successful project application. Energy Build 261:111946. https://doi.org/10.1016/J.ENBUILD.2022.111946
99. Satola D, Wiberg AH, Singh M, Babu S, James B, Dixit M, Sharston R, Grynberg Y, Gustavsen A (2022) Comparative review of international approaches to net-zero buildings: knowledge-sharing initiative to develop design strategies for greenhouse gas emissions reduction. Energy Sustain Dev 71:291–306. https://doi.org/10.1016/J.ESD.2022.10.005
100. Wuni IY, Shen GQP, Osei-Kyei R (2019) Scientometric review of global research trends on green buildings in construction journals from 1992 to 2018. Energy Build 190:69–85. https://doi.org/10.1016/J.ENBUILD.2019.02.010
101. Theodosiou TG, Papadopoulos AM (2008) The impact of thermal bridges on the energy demand of buildings with double brick wall constructions. Energy Build 40:2083–2089. https://doi.org/10.1016/J.ENBUILD.2008.06.006
102. Wei W, Skye HM (2021) Residential net-zero energy buildings: Review and perspective. Renew Sustain Energy Rev 142:110859. https://doi.org/10.1016/J.RSER.2021.110859
103. Zhang S, Wang K, Xu W, Iyer-Raniga U, Athienitis A, Ge H, Cho DW, Feng W, Okumiya M, Yoon G, Mazria E, Lyu Y (2021) Policy recommendations for the zero energy building promotion towards carbon neutral in Asia-Pacific Region. Energy Policy 159:112661. https://doi.org/10.1016/J.ENPOL.2021.112661
104. Lu Y, Zhang XP, Huang Z, Lu J, Wang D (2019) Impact of introducing penalty-cost on optimal design of renewable energy systems for net zero energy buildings. Appl Energy 235:106–116. https://doi.org/10.1016/J.APENERGY.2018.10.112
105. Chan AT, Yeung VCH (2005) Implementing building energy codes in Hong Kong: energy savings, environmental impacts and cost. Energy Build 37:631–642. https://doi.org/10.1016/J.ENBUILD.2004.09.018

106. Radhi H (2009) Can envelope codes reduce electricity and CO_2 emissions in different types of buildings in the hot climate of Bahrain? Energy 34:205–215. https://doi.org/10.1016/J.ENERGY.2008.12.006
107. Davies P, Osmani M (2011) Low carbon housing refurbishment challenges and incentives: architects' perspectives. Build Environ 46:1691–1698. https://doi.org/10.1016/J.BUILDENV.2011.02.011
108. Yuan B, Zhang Y (2020) Flexible environmental policy, technological innovation and sustainable development of China's industry: the moderating effect of environment regulatory enforcement. J Clean Prod 243:118543. https://doi.org/10.1016/J.JCLEPRO.2019.118543
109. Zahir MH, Irshad K, Shafiullah M, Ibrahim NI, Kausarul Islam AKM, Mohaisen KO, Sulaiman FAA (2023) Challenges of the application of PCMs to achieve zero energy buildings under hot weather conditions: a review. J Energy Storage 64:107156. https://doi.org/10.1016/J.EST.2023.107156
110. Ohene E, Chan APC, Darko A (2022) Review of global research advances towards net-zero emissions buildings. Energy Build 266:112142. https://doi.org/10.1016/J.ENBUILD.2022.112142
111. Van Nguyen M (2023) Drivers of innovation towards sustainable construction: a study in a developing country. J Build Eng 80:107970. https://doi.org/10.1016/J.JOBE.2023.107970
112. Ohene E, Chan APC, Darko A (2022) Prioritizing barriers and developing mitigation strategies toward net-zero carbon building sector. Build Environ 223:109437. https://doi.org/10.1016/J.BUILDENV.2022.109437
113. Ceglia F, Esposito P, Marrasso E, Sasso M (2020) From smart energy community to smart energy municipalities: literature review, agendas and pathways. J Clean Prod 254:120118. https://doi.org/10.1016/J.JCLEPRO.2020.120118
114. Rocha HRO, Fiorotti R, Louzada DM, Silvestre LJ, Celeste WC, Silva JAL (2024) Net zero energy cost building system design based on artificial intelligence. Appl Energy 355:122348. https://doi.org/10.1016/J.APENERGY.2023.122348
115. Siasiev A, Dudnikov S, Nikolaienko D, Hubal H, Bondarchuk O (2025) Evaluation of the effectiveness of decision-making algorithms in the management of renewable energy sources: a case study of Ukraine. Sustain Eng Innov 7:87–96. https://doi.org/10.37868/SEI.V7I1.ID415
116. Rocha HRO, Fiorotti R, Louzada DM, Silvestre LJ, Celeste WC, Silva JAL (2024) Net zero energy cost building system design based on artificial intelligence. Appl Energy 355. https://doi.org/10.1016/J.APENERGY.2023.122348
117. Elshafei G, Vilčeková S, Zeleňáková M, Negm AM (2021) An extensive study for a wide utilization of green architecture parameters in built environment based on genetic schemes. Buildings 11. https://doi.org/10.3390/BUILDINGS11110507
118. Flodberg K, Blomsterberg Å, Dubois MC (2012) Low-energy office buildings using existing technology: simulations with low internal heat gains. Int J Energy Environ Eng 3:1–9. https://doi.org/10.1186/2251-6832-3-19/FIGURES/9
119. Jovović I, Hrovatin N, Zorić J (2025) The role of energy performance certificates and energy-related financial literacy in choosing an energy-efficient home: evidence from Slovenian homeowners. Energy Effic 18:1–20. https://doi.org/10.1007/S12053-025-10320-6/TABLES/6
120. Kadhim N (2022) BIM and GIS data integration for the evaluation of building performance. Int Arch Photogramm, Remote Sens Spat Inf Sci-ISPRS Arch 46:135–139. https://doi.org/10.5194/ISPRS-ARCHIVES-XLVI-5-W1-2022-135-2022
121. Ou Y, Bailey N, McArthur DP, Zhao Q (2025) The price premium of residential energy performance certificates: a scoping review of the European literature. Energy Build 332. https://doi.org/10.1016/J.ENBUILD.2025.115377
122. Leskinen N, Vimpari J, Junnila S (2020) A review of the impact of green building certification on the cash flows and values of commercial properties. Sustainability 12:2729. https://doi.org/10.3390/SU12072729

123. What is an energy certificate? Construction how. https://constructionhow.com/what-is-an-energy-certificate/?utm_source=chatgpt.com. Accessed 23 June 2025
124. Warning over 'unreliable' energy bill codes that could end up costing you thousands of pounds|The sun. https://www.thesun.co.uk/money/28564734/warning-unreliable-energy-bill-codes-cost-thousands-epc-ratings/?utm_source=chatgpt.com. Accessed 23 June 2025
125. EPC benefits|EPCHome. https://epchome.co.uk/epc-benefits/?utm_source=chatgpt.com. Accessed 23 June 2025
126. Energy performance certificates (EPC): benefits and requirements. https://snapshot.ueniweb.com/blog/energy-performance-certificates-epc-benefits-and-requirements?utm_source=chatgpt.com. Accessed 23 June 2025

Chapter 3
Advances in Energy-Efficient Building Strategies

Abbreviations

BAPV	Building-Applied Photovoltaics
BEMS	Building Energy Management Systems
CCT	Correlated Color Temperature
CCTV	Closed-Circuit Television
DAP	Decoupled Adversarial Strategy
DRL	Deep Reinforcement Learning
GHG	Greenhouse Gas
HDR	High Dynamic Range
HID	High-Intensity Discharge
HVAC	Heating, Ventilation, and Air Conditioning
LCSs	Lighting Control Systems
LSTM	Long Short-Term Memory
MCS	Mechanically Controlled Shading Systems
MEMPs	Marine Equipment Manufacturing Facilities
PCM	Phase Change Materials
PV	Photovoltaic
RTE	Round-Trip Efficiency
YOLO	You Only Look Once

3.1 Introduction

The focus of this work is on the development and application of energy efficient strategies through built environment, solving several problems that are caused by increasing energy use and impact in the environment. The major goal is to investigate and develop creative methods that can improve building energy efficiency while

B. Duraković, *Advancing Zero Energy Buildings*, Green Energy and Technology,
https://doi.org/10.1007/978-3-032-16601-2_3

preserving occupant comfort and reducing the negative effects on the environment. For the purpose to lowering greenhouse gas emissions and saving operating costs, this chapter provides many recent developments in the field of energy efficient materials, systems and methods. Sustainable retrofitting methods, cutting-edge building materials, and many innovative technical interventions like smart control systems are just few of the subjects that are covered in this chapter. Using knowledge from diverse research, the study provides an effective structure that can help with applying energy efficient measures that are suited for various building types and climatic scenarios, into practice by using information from diverse research fields. A significant area of this research is the importance of moving towards sustainable building and urban planning methods to achieve international energy goals.

3.1.1 Energy Efficiency Strategies

Sustainable methods that manage the use of natural resources and maintain interior occupant comfort are crucial. Methods that applied to reducing energy consumption in existing and new buildings lead to mitigating environmental emissions. Therefore, changes and modifications made on existing buildings, that increase its energy efficiency, and improve its lifespan, are referred to as "retrofit", "retrofitting" and "refurbishment". This may include the use of contemporary technology to improve the use of resources or to improve building energy efficiency or provide greater occupant comfort [1, 2]. Typically, these terms are used to describe a building's partial or complete renovation and offer an opportunity to further improvements including sustainable practices [3, 4].

Reducing building energy consumption, cutting expenses, and mitigating the impact on the environment all depend on energy efficiency measures. Figure 3.1 shows fundamental energy efficiency strategies used in buildings. The strategies include high-performance insolation and building materials, energy-efficient windows and shading techniques, advanced Heating, Ventilation, and Air Conditioning (HVAC) system and controls, light efficiency and BEMS.

To increase heat flow resistance through the building envelope and lower energy loss, high-performance insulation is one of these strategies that is essential for strengthening building thermal performance. Windows are another important element in enhancing building buiding envelope and energy savings, which should be made to achlieve maximum natural light and minimize heat transmission. HVAC systems are essential for maintaining indoor comfort, but thanks to advances in energy-efficient technologies and better system management, they can lead to substential energy savings. Today, lighting efficiency is achieved through smart lighting systems, which is automatically adjusted based to occupancy needs and natural light levels. Lastly, a building energy use may be tracked, managed, and optimized in real time by integrating many technologies with BEMS. When combined, these strategies offer a comprehensive approach to enhancing energy efficiency in building.

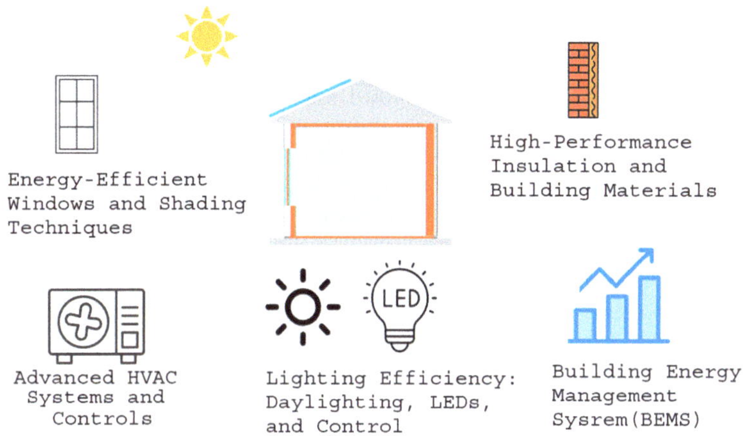

Fig. 3.1 Energy efficiency strategies

Based on the literature, Table 3.1 overview of the current challenges and future directions in various energy efficiency strategies. In the area of High-Performance Insulation, the primary challenge is the limited durability of bio-based materials, especially their susceptibility to water and fire damage. In order to improve performance and durability, future approaches will concentrate on creating hybrid materials and cutting-edge coatings.

The primary obstacles to energy-efficient windows are the high cost of installation and the issue of maximizing shading for different building orientations and climates. The focus of future research is on AI-driven adaptive systems that are capable of real-time adaptation to increase efficiency. The key challenge associated with HVAC systems is the lack of proper monitoring of environmental factors and changeable occupancy. One possible future approach is to combine AI predictive controls, like Deep Reinforcement Learning (DRL) systems, with renewable energy sources. The high costs of smart lighting are discouraging wider adoption for light efficiency. Low-cost sensors and streamlined manual control programs are considered as solutions to make smart lighting systems more accessible. Finally, complicated installations and a shortage of skilled workers are persistent problems in BEMS. To overcome these obstacles and increase the efficacy of BEMS, worker training programs and the simplification of AI and IoT integration are thought to be crucial.

Table 3.1 Challenges in energy efficiency strategies and future directions

Aspect	Current challenges	Future directions
High-performance insolation	Need for bio-based materials to be more water resistant and fire resistant due to their limited durability [5–7]	Research on hybrid materials and sophisticated coatings to improve performance and durability [5, 7]
Energy-efficient windows	Expensive installation; creates a challenge to optimize shade for many kinds of climates and building positions [8–10]	Creation of simulated and adaptive shading systems; application of AI for real-time modification [8, 11, 12]
HVAC system	Poor supervision of variable levels of occupancy and environmental conditions [13, 14]	Combining renewable energy sources with the AI predictive controls (like DRL) [15, 16]
Light efficiency	Smart lighting system prices are high, it turns off the consumer [17, 18]	Creating low-cost sensors and development of simple programs for manual control [19, 20]
Building energy management systems (BEMS)	Installation and repairs are complicated, there is an absence of trained workers to operate [21, 22]	Simplification of AI and IoT systems; training initiatives for employees [21, 22]

3.2 Thermal Insulation and Building Materials

3.2.1 Hemp Stalk-Based Insulation

Insolation that is made from hemp stalk is a popular, high-performing, environmentally friendly building material. The thermal conductivity of hemp stalk, which is sometimes considered as agricultural waste, ranges from 0.049 to 0.082 W/mK [5]. According to industrial standards, the material is lightweight, and it provides great thermal and acoustic insolation qualities. This material has great acoustic absorption (0.88–0.95 dB/dB) and it performs similarly to more traditional materials, such as expanded polystyrene. To improve performance of this material, it is bonded with recycled cardboard fibers, and citric acid treatment that improves cellulose crosslinking [5].

Eco-friendly coating, such as colophony and gum arabic, are also utilized to improve resistance against moisture and fire. Mechanical testing shows that hemp-based materials are tough and competitive with traditional insulation, while gum arabic coating further ensures fire safety compliance. Hemp-based insulation supports sustainable construction practices by reducing carbon emissions-it may store up to 325 kg of CO_2 per ton of hemp stalk and encourage circular economy because it is recyclable [6]. Research is ongoing to enhance these materials long-term durability, explore new coatings, and boost their fire and water resistance, making them a more appealing choice for green building projects, even though they already outperform traditional alternatives in terms of sustainability and insulation efficiency [7].

3.2.2 *Mycelium-Based Composite Insulation (MBC)*

Mycelium-based composites (MBCs) are a new and environmentally friendly insulating material that combines biological waste into a porous, lightweight structure, and is produced from fungus mycelium. They offer a sustainable alternative to traditional insulating materials like polyurethane foam because of their greater thermal properties and lower environmental impact. MBCs compete with other insulators such as expanded vermiculite and lightweight expanded clay aggregate due to their good thermal conductivity of 0.069 W/mK. As these materials are renewable, biodegradable, and require little energy to produce, they align with principles of the circular economy. MBCs, which have an optima thickness of 0.08 m, reduce energy consumption by preserving indoor temperature while balancing cost and performance. These characteristics make mycelium-based composites a good choice for green construction, particularly in regions with significant temperature fluctuations [23].

Energy Storage within Phase Change Materials (PCM) in building walls are very popular options for high energy performance building by improving their energy efficiency and lowering carbon emissions. Thermal energy is stored/released in the form of latent heat during the PCM phase transition withing a narrow temperature range. By absorbing excess heat during warmer periods and releasing it during cooler times, PCMs help maintain indoor thermal comfort, thereby reducing the reliance on heating and cooling systems. This reduction in energy consumption leads to lower carbon emissions, supporting sustainability goals. In this process, the thermal inertia of building envelopes is significantly improved, and the indoor temperature fluctuation is significantly reduced. Table 3.2 shows key results on the application of PCMs in construction materials.

PCM walls can keep stable interior temperatures for a longer period. They include a 0.8-h reduction in temperature delay and a 0.6 °C drop in inner surface temperature. By reducing yearly energy that is spent on cooling, by 7.6% and carbon dioxide emissions by 2% in buildings that use them, they meet the carbon neutrality standards [24]. For example, it was also discovered that "sandwich designs" with minimal phase-change locations are also very good at minimizing the heat transfer.

What is very important is to place PCM close as much as possible to the interior side of the wall. Research also shows that PCM that are placed farther from the exterior really reduce the heat flow and improve internal temperature regulation [28, 29]. A lot of studies were done on the thermal efficiency of PCM walls, and these studies have shown that PCM walls can reduce heat transfer, improve wall insolation and reduce the interior temperature fluctuation. Also, according to those studies, PCM walls have even more consistent temperature ranges than those typical walls, because their temperature oscillate between 21.2 and 35.8 °C [24].

Table 3.2 PCM application in construction materials

Area of application of PCM	Results	References
Building walls (hemp-based insolation)	Thermal conductivity: 0.049–0.082 W/mK. Reduces CO_2 by up to 325 kg/ton. Promotes circular economy	[5–7]
Concrete (phase-change energy storage)	Decreases cooling energy by 7.6%. Reduces CO_2 emissions by 2%. Improves indoor thermal stability	[5, 24–26]
Gypsum board (mycelium-based composite)	Thermal conductivity: 0.069 W/mK, eco-friendly, and uses less energy by keeping temperatures constant	[23]
Brick (bio-based materials)	Climate-specific adjustment lowers heating and cooling demands and keep environmental sustainability	[27]

The integration of PCM into walls contributes to solving a big building energy problem, such as minimizing energy losses and demand for HVAC systems. PCM that were constructed in colder areas proved that they improve thermal stability and, they can control indoor temperatures more effectively than regular insolation methods. Buildings that have PCM walls save on heating and cooling energy, which automatically reduce carbon footprint and reduce their need for fossil fuels [25, 26].

To reduce building energy consumption and reduce environmental impact, a high-performance insolation and energy storage materials will play a significant role. Heating and cooling systems consume nearly 40% of building energy use, therefore it is important to improve energy efficiency of building envelopes by applying thermal insulation techniques [27]. Because of that we have a need for different eco-friendly materials, artificial composites and other techniques and processes that can reduce the need for energy consumption. Accordingly, some bio-based materials like hemp, flax, sisal fiber and so on, have become more popular.

Sisal fiber is environmentally friendly and has great thermal conductivity of 0.038 W/mK. Although synthetic materials have great thermal resistance, some materials like aerogels, expanded polystyrene and extruded polystyrene are not usually that much used because they have relatively high production coast and also environmental concerns [27].

Double-wall masonry buildings with inner air layers act as thermal buffers, they also can reduce the heat transfer which can improve the effectiveness of insulation. So, by decreasing the insolation thickness and change of air speed, techniques such as air gap within insulation layer additionally prevent heat losses. A good example is Morocco, where the air-gap solution saved 3.72% in the summertime and 3.68% in the wintertime, by lowering the insolation thickness from 7 to 5 cm. The study showed the importance of specific insolating techniques that can reduce energy consumption [27].

Furthermore, the environmental benefits of high-performance insulation materials cannot be underestimated. Bio-based materials, produced from renewable resources, significantly decrease the total energy used and carbon footprint of construction when compared to synthetic materials such as polystyrene and polyurethane foams. By combining sustainable materials with innovative construction techniques, buildings can achieve increased thermal comfort and energy efficiency while helping achieve global sustainability goals. For example, sisal fiber-based Moroccan buildings envelopes increase thermal insulation and satisfy the county's new thermal regulations, which are intended to reduce energy consumption and boost construction industry energy efficiency [27, 30, 31].

3.3 Windows and Shading Techniques

Shadowing from nearby buildings, trees, or architectural features like overhangs and protrusions is a significant barrier to optimizing energy efficiency in dense urban settings. Partial shade, which may vary during the day and season, might make it more difficult to maintain consistent energy performance. Buildings must have energy-efficient windows and shading techniques to balance solar heat gain and improve indoor comfort. PCM walls are commonly designed using materials that improve temperature regulation and thermal resistance, such as gypsum-based PCMs and porous expanded perlite. One of the most important components of renewable energy generation in urban environments is Building Applied Photovoltaic (BAPV) systems, which take advantage of unused rooftop space to generate power for on-site use or grid contribution without requiring additional land.

These systems require only solar energy, they are economical and minimize transmission losses, they are efficient and beneficial for the environment. On the other hand, these systems have some drawbacks related to the surrounding environment they are installed. Trees, clouds or architectural elements may create shading, which can affect its efficiency by lowering energy output. The impact of shading can be lessened by bypass diodes and power optimizers, but these solutions have disadvantages such as voltage imbalances, complex systems, and higher maintenance costs. Advanced techniques like dynamic reconfiguration employing real-time optimization algorithms and static reconfiguration improve the production of energy by handling shading constantly and adaptively [32]. Despite being more costly and complex, innovative methods provide more flexibility and efficiency, making them especially useful in scenarios that are prone to shading. When combined with such innovative strategies, BAPV systems are crucial to sustainable building since they maximize urban energy output and lessen reliance on non-renewable resources [33].

Energy-efficient windows and shading systems are crucial to solving the increasing energy demands and environmental issues associated with buildings. Between 1990 and 2019, global building CO_2 emissions rose by 50%, while global energy consumption rose by 38%. Among these, non-residential structures increased by a significant 54%. In particular, the sharp 161% increase in electricity consumption highlights how urgently energy-efficient alternatives are needed. Building envelopes, which act as protective barriers regulating the circulation of energy between interior and external spaces, are necessary to optimize energy use. They significantly affect thermal efficiency, which is impacted by several factors, including climate, design variables, material properties, and building occupancy dynamics [34]. The demand for high-end office environments that prioritize health, and efficiency has led to improvements in building envelope designs. However, creating a high-performance facade is challenging, especially if there are large windows that enhance daylighting and views while also boosting summer heat intake and reducing winter heat storage. To address these problems, shading solutions play a significant role in reducing peak heat loads and the need for cooling. Advanced shading skins not only provide the standard thermal benefits but also enhance external visual connectivity, which aligns with modern environmental and aesthetic goals [34].

Due to growing economic and environmental concerns about energy use, research over the past few decades has concentrated on improving the energy efficiency of residential and commercial buildings. The building envelope has a significant impact on productivity, health, and quality of life. Windows and shading systems are crucial for balancing energy efficiency with the thermal and visual comfort of people [8]. Windows are essential to a building's energy efficiency because they allow for natural ventilation, heat transmission, and solar radiation absorption. Compared to other building envelope components, they have an approximately five times greater impact on energy efficiency. However, windows without shading mechanisms can lead to other issues like glare, overheating, and increased energy costs for cooling. Solar control systems with shading devices provide an effective way to reduce heating and cooling energy consumption by regulating sunlight entry to prevent overheating in the summer and allowing sunlight to enter in the winter. Appropriate shade designs enhance human comfort and energy efficiency by improving thermal and visual conditions [8]. To increase the energy efficiency of older buildings, shading systems may also be modified.

New HVAC systems and state-of-the-art technologies like big data analytics, smart sensors, and machine learning methods are commonly used together with windows and shading systems. The benefits of shade systems have been shown in a range of architectural contexts and climates. In warm areas, energy use is significantly reduced by carefully chosen window designs and optimal shade patterns [8]. Adding horizontal external shading along with specific HVAC temperature set points may save up to 46% on energy costs in a historic building in Egypt [10]. In the same way, research from the US, China, and Oman showed that external shading might save energy by 16.56–40%. Energy savings of over 40% have also been shown by recent modular dynamic facades designed for hotter climates like Iran. The geographical adaptability of these techniques was highlighted by the discovery that shade devices were essential for controlling daylight access [8].

Horizontal shade systems generally perform better than vertical ones, and flexible designs that adjusted to specific facade orientations and climatic circumstances provide the best results. For instance, it has been shown that using the right shade slats and angles can cut down on glare and improve daylight performance by up to 44%. Additionally, integrated shading systems, such as expanded-metal shading in Japan or multi-layer blind systems in Tehran, made significant improvements in daylight measurements while reducing annual sunshine exposure [8]. Horizontal shading at slight upward or downward preferences maximizes energy efficiency by reducing cooling loads while improving natural lighting. Multi-slat layouts are beneficial for daylighting and energy conservation, but they must be balanced correctly to avoid glare or discomfort from heat. Sharply angled shading systems, for example, performed better in two-window configurations, improving lighting by 59.60% and reducing energy consumption intensity by up to 31.29% [8].

Smart shading systems can adapt to shifting environmental circumstances and optimize energy efficiency and comfort for users with the aid of sensors and programmed controls. Shade adjustments have been adjusted with machine learning techniques to maximize sunshine penetration and reduce glare. In order to balance thermal comfort, visual comfort, and energy economy, architects can do multi-objective optimization using simulation tools [8]. Several case studies illustrate the value of shade systems in different building types and climates. For instance, two office room designs in Tehran were analyzed: one that used a single southern window and the other that had windows on multiple faces. The best combination between comfort for users and energy efficiency was found to be provided by horizontal shadow angles up to 20°, with daylight benefits of 12.1% and energy reductions ranging from 7.9 to 29%. In multi-facade arrangements, combining horizontal and vertical shade systems led to a 59.60% rise in daylight measurements and a 31.29% reduction in energy usage [8–10, 35].

Mechanically controlled shading systems (MCS) go beyond shading by including dynamic control techniques to automatically improve energy performance. Some of the many developments in MCS are artificial neural networks that are used for controlling shades, also solar tracking louvre systems, and integration with external and internal shades. Comparing these systems with traditional fixed shading it is observed that these systems provide energy savings up to 49.7% [11].

Thermochromic windows have been demonstrated to reduce energy consumption by selecting the appropriate phase transition temperatures for specific regions. However, there are still problems, such as reduced visible light transmittance after the state shift, which could increase the energy needs for artificial lighting. One modern innovation that shows promise for boosting energy efficiency while maintaining adequate daylight levels is thermochromic hydrogel wood overlays for windows. New developments in adjustable shading systems systems are expected to get over these challenges by independently modifying near infrared and visible light to improve building energy efficiency [11]. There are still numerous challenges to be solved despite improvements in shading technology, especially when it comes to maximizing shade strategies for a range of building types and climates. Most of the current research focusses on verifying the energy-saving advantages of specific shading strategies rather than providing complete design frameworks for practical applications. The need for uniform decision-making tools that consider different shading approaches, and their complimentary effects is highlighted by this difference [11]. Research indicates that the possible energy savings of shading systems vary significantly depending on the environment and building characteristics. For shading devices like thermochromic adaptive windows, cooling-excited tracking louvres, tracking louvres, and static louvres, for example, studies have modified the window-to-wall ratio, slat width, and phase transition temperature. Researchers have used computational tools such as EnergyPlus and GenOpt to do multi-parameter optimization to figure out optimal shade arrangements for different climates [11, 12, 36, 37].

Integrating PCMs into window systems offers a promising approach to enhancing building energy efficiency. PCMs absorb and release thermal energy during their phase transitions, helping to modulate indoor temperatures and reduce reliance on heating and cooling systems [38, 39].

3.4 HVAC Systems and Controls

Today's technology is improving rapidly and there are bigger requirements for improved HVAC. The traditional HVAC systems can no longer keep up with those standards and provide energy saving and comfort. Therefore, it is required that modern buildings incorporate energy-efficient HVAC systems, not only to support sustainable energy practices but also to enhance indoor environmental quality. One of the new technologies that helps in optimizing air quality, occupant comfort and energy usage, is the implementation of DRL within HVAC automation through the use of neural networks [30].

Therefore, researchers developed a technique that is called Decoupled Adversarial Strategy (DAP), so this technique stimulates problems to appear during straining, because HVAC systems can analyze the time-based data more, they use Long Short-Term Memory (LSTM) networks. However, the system can be trained to anticipate and address these issues in the future. And finally, the combination of DAP and LSTM-enhanced DRL can help with this topic around HVAC systems, using less energy but still providing the same comfort that is required.

HVAC plays an important role for thermal comfort, Indoor Air Quality and energy efficiency. Every technique has its own advantages and disadvantages. By incorporating advanced ventilation systems, energy expenses in new construction can be reduced by up to 60% while keeping superior air quality and occupant comfort. Therefore, advanced HVAC systems are crucial for environmentally friendly and energy-efficient construction designs [40].

HVAC systems can be synchronized with shading techniques to enhance natural lighting and reduce heat gain, which decreases the overall energy load [41]. In the context of windows that are energy-efficient and shading techniques, the results emphasize the necessity of combining predicted AI models with shading techniques to improve thermal and lighting efficiency. An Adaptive Neuro-Fuzzy Inference System as a forecasting tool was introduced demonstrating its ability to manage complex, irregular relationships between weather and HVAC outputs. This approach reduces energy consumption and ensures indoor thermal comfort, thereby aligning with the goals of energy-efficient window and shading system designs [41]. It is observed that ventilation and heat load efficiency are greatly affected by wind speed, relative humidity, dry bulb temperature, and other environmental conditions. Ideal conditions, which are defined as 45% relative humidity, 13 °C dry bulb temperature, and 5 km/h wind speed, lead to major efficiency increases compared to pre-COVID-19 guidelines, with ventilation rates reaching 93% and heat load optimization at 97%.

Shading devices, such as adaptive or thermochromic systems, respond continuously to changes in their environment, much like the predicted AI algorithms. These systems can increase indoor temperatures while reducing the need for air conditioners by modifying solar heat gain and absorption. For instance, mechanically controlled shading systems such as thermochromic windows and louvres that adapt to the quantity of solar radiation can enhance HVAC performance by lowering excessive heat loads and maintaining visual comfort. By integrating such shading approaches with AI-driven technology, buildings can achieve a better balance between energy efficiency and resident comfort [41]. Additionally, advanced HVAC systems could assist in promoting improved air circulation, filtration, and respect the WHO indoor air quality regulations. Based on real-time environmental data, accurate heat and ventilation rate adjustments are made possible using AI models. Similarly, to maximize natural ventilation, minimize solar heat gain, and satisfy specific indoor air quality requirements, shade systems could be used in conjunction with HVAC controls [41–44].

HVAC systems are substantial electricity users, accounting for over 50% of power demand. Therefore, it is important to build state-of-the-art, eco-friendly HVAC systems that prioritize energy efficiency and sustainability. Enhancing HVAC efficiency can be effectively achieved through the integration of renewable energy sources, with solar power emerging as one of the most promising options due to its sustainability and eco-friendly nature. Solar powered HVAC system usually uses photovoltaic (PV) panels and solar thermal collectors. PV panels produce electricity that is used to power HVAC systems. To improve the efficiency of PV panels, some advanced cooling techniques, such as nanofluids, can be used. Hybrid nanofluids, which are composed of a mixture of different nanoparticles, have proven to be highly effective in enhancing heat transfer performance, with reported efficiency gains of up to 36.67%, making them promising candidates for applications in solar thermal collectors and cooling systems [15, 45, 46].

HVAC system efficiency remains a critical concern, as a significant portion of energy is often wasted due to operational faults such as pipe corrosion, compressor failures, sensor malfunctions, and refrigerant leaks. These issues harm thermal comfort and indoor air quality in addition to reducing equipment life. Building systems with poor control methods might lose 15–30% of their energy. To improve system efficiency and decrease the waste, engagement of rapid Failure Detection and Diagnosis techniques is essential, enabling early identification and resolution of faults to enhance system reliability and performance [13]. The Varible Refrigerant Flow systems are becomeing popular due to affordable investment costs, easy instalation and ability to adjust refrigerant flow. Learning models like convolutional neural networks and decision trees have high procent of accuracy, 97%, since they are one of the best options in identifying problems like refrigerent leakage [13].

The best option taht can help to detect any defect in early stage is predictive maintenance to prevent significant failures. This method may use cloud-based machine learning, and it is suitable for HVAC systems that are already in place, and it is also functioanl but in the other hand economical. Main focus is on refrigeratant leaks, but in the future it can also expand and try to sole problems with engine and airflow blockages [13, 14, 47–49].

Metrics such as Round-Trip Efficiency (RTE) and energy storage capacity are used to measure the load flexibility of HVAC systems. Because load shifting can also cause heat loss during precooling, and it can increase the consumption of energy. Insulation and thermal mass have also a big impact on RTE values, they go from 0.16 to 1.3. Model Predictive Control is one of the advanced control systems, which improves energy flexibility by changing HVAC operations depending on environmental and demand conditions [50–52]. Researchers are increasingly recognizing the importance of non-stationary and unpredictable modeling techniques to accurately capture such unpredictability. Incorporating non-stationary patterns into HVAC energy consumption models improve prediction accuracy, thermal comfort, and save energy [16, 53, 54].

3.5 Lighting System

Lighting systems are a critical component for improving energy efficiency, as they contribute 15–20% of global energy consumption and 5% of Greenhouse Gas (GHG) emissions. For commercial and office buildings, this share can be higher and reach up to 20–45%, while in some sectors, such as retail, lighting can account for up to 80% of total energy consumption [55, 56]. Around 50% of the energy used in educational buildings is used in classrooms [57]. Therefore, energy-efficient lighting solutions, such as LEDs, daylighting, and smart controls, are very important for reducing energy consumption in buildings, costa and GHG emissions. It is known that LEDs are the best option for long lifespan as well as good economical option [17]. Daylighting, which allows natural light to enter spaces through skylights, not only reduce emissions, costs, energy consumption, but it increases worker comfort and increase in productivity for 5–16% [17].

Lighting systems are a good place to start with making changes to energy management because due to its simplicity to change and upgrade. For example, LEDs and daylighting can be used to replace fluorescent and High-Intensity Discharge (HID) lights, saving quite a bit of energy without interfering with business operations. Another crucial aspect is behavior modification, since energy-efficient staff practices can result in additional savings of up to 19% [17]. An example from a Turkish Marine Equipment Manufacturing Plant (MEMP) reveals how a thorough approach to lighting efficiency may encourage cleaner production and increase productivity. The factory, which uses 109,169 kWh for lighting yearly, generated significant energy savings through daylighting, behavior modification, and energy efficiency improvement. The integration of these methods also addresses broader industrial issues, such as rising energy costs and environmental types, which are especially significant for energy-intensive businesses like MEMPs. By conducting detailed lighting energy audits and combining behavior-focused strategies with technological advancements, industrial facilities may significantly reduce energy usage, save expenses, and contribute to global sustainability goals [17, 58].

Lighting efficiency has become a critical issue in educational settings due to its impact on the use of energy, student happiness, and visual comfort. Improved Lighting Control Systems (LCSs), daylighting, and LEDs are only a few of the methods utilized in contemporary classrooms to provide enough illumination and improve energy efficiency. The efficiency, brightness level adjustability, and flexibility to decreasing and switching—all of which are vital for dynamic lighting situations—make LEDs an essential part of modern lighting systems, even though this is never directly mentioned. In addition to artificial lighting, daylighting strategies such as shade controls and window blinds optimize natural light use while reducing discomfort and glare [18].

LCSs are an essential component of energy efficiency. For various activities, these devices constantly change the lighting conditions. They frequently have sensors to detect motion, light, or occupancy. For example, Shi et al. developed an LCS with automated, timed, and manual modes to control light intensity between 100 and 300 lx, while Luansheng et al. introduced a system with light sensors to maintain light levels between 200 and 500 lx by switching on lamps only in designated locations. De Rubeis et al. and Martirano developed LCS technology and obtained energy savings of up to 50% and 69.6%, respectively, by combining motion and sunshine sensors [18]. Lee et al. showed a context-aware LED lighting control system that use occupancy detectors to change lighting according to activities, allowing them to monitor the movements and conduct of teachers and students. Budhiyanto and Chiou demonstrated an LCS prototype that, based on how the classroom is organized and utilized, can reduce energy use by 20–80% [18]. To address the high cost of sensors, digital cameras have been suggested as an affordable alternative for detecting luminance and illuminance. These cameras use High Dynamic Range (HDR) photography to capture high-resolution luminance data over a wide field of view. Although it is slightly less accurate than standard luminance meters, this method is frequently used for lighting assessments and has shown strong correlations with subjective glare scores. In one study, an HDR camera system was utilized to maintain visual enjoyment for 88% of working hours, maximize daylighting, and decrease energy costs by roughly 48%. Despite their efficacy, automated systems commonly face difficulties since residents typically overcome controls to personalize lighting to their preferences [18]. Utilizing this technology has resulted in considerable energy savings. De Rubeis et al. found a 69.6% decrease in energy use, Martirano achieved 50%, while Budhiyanto and Chiou claimed savings of 20–80%. However, there are certain disadvantages to these systems, including the need for dynamic adjustments to fit a range of educational activities, the costly initial sensor installation, and the need for human control actions to increase user satisfaction [18]. The HDRi Surveillance Lighting and Daylighting Control System is a significant innovation in this area. This system uses commercial security cameras with HDR vision sensors to automatically manage the lighting in classrooms [18, 19].

This combination is very important when considering both new and old buildings, also including historic buildings, which can cause a lot of problems for this topic. They have huge historical significance, but their refurbishing options like wiring and structural changes are very restricted. Non-intrusive alternatives, including inner roller blinds and smart lights that can be controlled wirelessly, offer a helpful option in these circumstances. These systems maintain the original architecture of these buildings while also providing improvements in lighting efficiency and significant energy savings. When compared to non-integrated systems, properly constructed combined daylighting and lighting control systems can save energy expenses by 35–60% [20]. By replacing traditional fluorescent lighting with energy-efficient LED lighting, energy consumption can be further reduced by about 50%. Dynamic lighting management systems can adjust the Correlated Color Temperature (CCT) and light intensity to suit user needs and environmental conditions by changing the warmth or coolness of LED lighting. This adaptability increases occupant well-being and facilitates the development of integrated lighting systems, commonly referred to as human-centric lighting, in addition to encouraging energy efficiency. Through the dynamic modification of intensity and CCT to align with circadian cycles, these systems increase wellbeing and productivity. Residents like the dynamic features of these systems, despite the fact that there is currently no evidence of circadian benefits in real-world applications [20]. Shade adjustments and energy-efficient lighting are also crucial for enhancing visual comfort and saving electricity. Interior roller blinds are a good way to reduce solar radiation and prevent eye strain from too much light. Flexible LED lighting and motorized roller blinds are two examples of automated lighting and shading systems that can successfully react to shifting environmental conditions. However, these systems can be challenging to set up and order, particularly in older buildings with complex geometry and limitations. Wirelessly connected systems are commonly utilized as a solution since standalone sensing devices consume a lot of energy, however they may not always be the best choice [20]. Despite the advantages of automated methods, problems remain. Human control is preferred by many users, especially in personal circumstances. People may override or even harm automated systems if they are not properly equipped or do not have a manual override option. Incorrectly constructed controls may also lead to a decrease in user awareness, motivation, and satisfaction. These issues are made worse by architectural limitations that make system set up and operation more difficult in older buildings [20]. Manual systems might offer an effective alternative when passenger behavior is considered at every stage of the design process. Research shows that many lighting and shade changes occur as people arrive or leave and are rarely changed during work hours, unless they are triggered by outside causes like glare or bright sunlight. Some strategies to encourage energy-efficient behavior include placing controls close to work areas, designing simple user interfaces, and setting default settings, such as drawing the shades and shutting off the lights at the end of the day. Behavioral interventions including training, urging, and user interaction can further enhance the efficiency of manual systems [20]. Research proved the effectiveness of a low-cost, quiet Shading

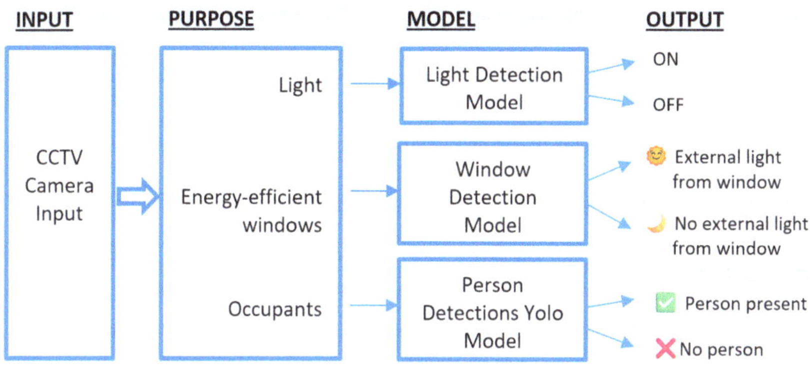

Fig. 3.2 Graphical representation of different object detection model performances

and Electric Lighting System in a historic building's office. The system's motorized inside roller blinds and changeable, adjustable lights were controlled wirelessly by desk-mounted remote controllers. Behavioral interventions were incorporated into the system to maximize energy efficiency. Default settings ensured that the lights were turned off at the end of the day and that the shades were lowered to promote energy-efficient behavior. During the 12-week observation period, which involved six users, the system showed significant energy savings while maintaining high levels of user acceptance and satisfaction [20, 59, 60]. A comparative overview of three common object detection models including person detection, window detection and the light detection model under different lighting conditions is shown in Fig. 3.2.

YOLO (You Only Look Once) object detection model that is specifically trained or used to detect people (persons) in images or video frames. Person detection is the task of identifying and locating people in an image or video. During the day it works the best, but over the night it does not work. Window Detection Model work well during the day providing a high visibility. Windows are usually clear in daylight because of contrast, edges, and reflections. At night there is low or no light presence providing poor visibility. If there is no artificial lighting, windows may blend in with the wall or appear as dark rectangles, which may confuse the model. Light Detection Model detects that the average brightness exceeds the set threshold, it concludes that the light is ON and may adjust the room's lighting accordingly. When the average brightness falls below the threshold, the system determines that the light is OFF and could trigger actions such as turning the lights ON for improved visibility. Table 3.3 provides a comparative summary of the three models.

It has been demonstrated that using LED lighting is a very effective energy-saving strategy. Even though T5 lights may occasionally work better when dimmed, LEDs use less energy than traditional T5 bulbs. For example, T5 luminaires usually save 77.7% on energy expenses, while LED DC supply luminaires can save up to 93.3%. This illustrates how LED technology offers significant advantages in terms of reduced use of energy [57]. Utilizing sunshine and daylighting are also crucial

Table 3.3 Comparison of light detection, person detection, window detection models, under different lighting conditions

Model/condition	Person detection (YOLO)	Window detection model	Light detection model
Purpose	Detect humans in images/videos	Detect window locations/shapes	Detect whether artificial lights are ON or OFF
Input	Image or video frame	Image or video frame	Image or video frame
Output	Bounding boxes for people	Bounding boxes/masks for windows	Binary: light ON or OFF
Use cases	Smart buildings, surveillance	Smart buildings, layout detection	Smart homes, energy automation
Works in total darkness	No—can't see people	No—windows not visible	Maybe—if artificial light visible in camera
Works in low indoor light	Degraded accuracy	Inconsistent—shadows can confuse	Yes—with good tuning
Works in daylight	Yes—performs best	Yes—windows clearly visible	Maybe—sunlight can be mistaken for artificial light
Lighting sensitivity	Moderate	High	Very high

components of energy efficiency in school buildings. When natural light is used effectively, shading solutions can reduce glare, increase eye comfort, and significantly reduce energy consumption. Photosensor-based daylight harvesting systems are promisable at lowering energy consumption, which is in keeping with the goals of Near Zero Energy Buildings. For example, the general daylight control system with occupancy sensors developed by Tullio De Rubeis et al. reduced energy expenses by up to 69.6%. While Delvaeye et al. found yearly energy savings of 18–46% with the use of daylight management systems, Yu et al.'s analyses for educational buildings in the UK predicted yearly energy savings of 40–46% by daylight utilization [57]. Energy efficiency is further increased by using LED lighting with advanced lighting controls, such as photosensor-based dimming systems. Simulations using DAYSIM software and controlled measurements showed the possibility of significant savings. For example, using LED DC supply luminaires with photosensors led to monthly energy savings of up to 94.6%. These technologies enhance the functionality and adaptability of lighting systems in educational settings while also improving energy efficiency [57].

Lighting energy efficiency is also greatly influenced by initial design and consumer behavior. Many schools have over-illumination due to poor lighting design, which increases installed power and energy consumption. Additionally, human behavior, such as leaving lights on for more time than necessary, may outweigh the benefits of energy-efficient devices. For example, it was shown that a 22% rise in electricity use in new Canadian schools was caused by occupant behavior. These behavioral difficulties must be addressed to maximize energy savings [57]. Two international efforts to reduce energy consumption in school buildings are new projects and the use of advanced technologies. For example, the renovation of the Tito Maccio Plauto building reduced primary energy use by 81%. Improvements to building envelopes, mechanical, electrical and plumbing services, along with the setup of building management systems and solar energy, resulted in primary energy savings of 29.87 and 38.29% for two Tehrani school buildings. Despite Greece's abundant sunshine, daylight harvesting is still not commonly used, resulting in a significant energy waste in educational buildings [57, 61–63].

The shipping industry is a key contributor to fuel consumption, which is responsible for 24% of GHG emissions and one-third of global energy consumption. Despite the focus on reducing emissions during the ship's operational phase, the manufacture stage of MEMPs is energy-intensive and significantly contributes to environmental effects. To create essential ship components like propellers, rudders, and deck equipment, MEMPs employ energy-intensive processes like melting, cutting, polishing, and heat treatment [56]. Energy efficiency requires behavior change in addition to technology solutions. By taking simple steps like turning off unnecessary lights, employees can reduce their energy consumption by up to 19%. To encourage an energy-conscious culture, however, energy conservation must be incorporated into workplace policies and performance reviews. Alongside technology improvements, behavior change is crucial to achieving the full potential of energy savings in MEMPs [56]. Rising energy prices motivated on by geopolitical issues, like restrictions related to the war in Ukraine, make it harder for MEMPs in Europe to maintain economic viability and boost competitiveness. Since these facilities are governed by the Paris Agreement rather than maritime norms, they must undergo a green transition to reduce energy consumption and emissions [56]. Lighting in MEMPs and other manufacturing industries uses about 16% of global energy. An important and cost-effective strategy is to convert to energy-efficient lighting, such as LED lights [56].

Lighting energy audits revealed that the machine shop's fluorescent tubes consumed the remaining 109,169 kWh of energy each year, while the foundry's HID lights consumed 96,854.4 kWh. By employing natural light and energy-efficient lighting solutions and maintaining optimal lighting conditions through simulations using tools such as DIALux, the facility found significant energy-saving potentials Beyond monetary gains, MEMP lighting energy efficiency is essential. It addresses broader concerns including respecting the Paris Agreement and reducing environmental impacts. Despite the lack of study on energy-efficient lighting at the industrial scale, recent studies show the revolutionary potential of natural light and LED lighting in industrial buildings. Combining intelligent control technology and promoting behavior change results in a complete strategy to increase lighting energy efficiency, allowing MEMPs to save costs, increase competitiveness, and promote global sustainability goals [56, 64–67].

3.5.1 Building Energy Management Systems

BEMS integrate various components to monitor and optimize energy usage within buildings, aiming to enhance efficiency and sustainability. Therefore, modern buildings must have BEMS to achieve energy efficiency and reduce environmental impacts. Key elements of a BEMS include sensors and meters that collect real-time data on parameters such as temperature, humidity, occupancy, and lighting levels. The date is transferred to a centralized management platform for processing to identify inefficiencies and suggests optimization strategies. Based on these insights, controllers and actuators adjust HVAC system or lighting to align with energy-saving objectives [21, 68].

By combining intelligent outstations and central control stations, BEMS makes it possible for dynamic monitoring, real-time data collection, and building operations management. This enables specialists and site managers to identify mistakes, such as broken equipment or incorrectly set-up systems, and take corrective actions to improve energy efficiency [21]. One of the main strategies for energy efficiency that BEMS allows is tracking and focused on energy usage. Systems track and analyze power and fuel consumption to find efficiency. For example, BEMS can spot operational problems like heating system improperly positioned burner nozzles, which can save a lot of energy when fixed. BEMS additionally improves heating and cooling schedules, ventilation rates, and lighting controls to maintain comfortable rooms with the least amount of energy [21].

BEMS are very good at managing HVAC systems, especially by utilizing energy-efficient solutions such as displacement ventilation, chilled ceilings, and mixed-mode systems that combine natural and mechanical ventilation. While improving strategies for natural ventilation like stack and wind-driven airflow through automation, BEMS reduces the demand for mechanical cooling while keeping good indoor air quality. These systems especially work effectively by lowering energy consumption during climatic and seasonal changes. Lighting management forms another important area in which energy efficiency is contributed by BEMS. Combination of solar energy with user responsive lighting controls allows the BEMS to reduce energy usage by adjusting artificial lighting as per the availability of natural light and presence of the users. Flexible LED lighting solutions can double efficiency by allowing color temperature and light intensity adjustment according to user preferences and environmental conditions. BEMS automatically adjusts the building operation variables against changes in conditions, using advanced techniques for gain, like fuzzy logic and PID control. For example, airflow and temperature setpoints can be continuously adjusted based on usage trends, external weather conditions, and energy consumption targets. Additionally, BEMS uses digital communication technologies such as LANs and internet-based tools to allow remote monitoring and control, therefore improving flexibility and efficiency [21].

False sensors, also complicated installation and the requirement for user training are only some of the disadvantages of BEMS. But system performance is also improved by using proper licensing and input from building occupants. According to some studies that were done, BEMS can drastically cut energy use. Only thermostat adjustment can save up to 14%. They used simulations that took temperature and space size into consideration. For example, community centers installed BEAMS and improved insulation, decreased its use of natural gas by 62% and electrical energy by 56% [22].

Recent advancements in artificial intelligence (AI) and smart technologies have greatly increased the potential of BEMS. AI techniques like deep learning and fuzzy logic are used to optimize lighting and HVAC systems in real-time while balancing energy savings and occupant comfort. Given population data and weather projections, smart control algorithms automatically adjust energy settings to reduce waste and increase efficiency. Smart thermostats connected with BEMS further enhance performance by considering factors such as outdoor weather, building envelope condition, and room utilization [22].

3.6 Remarks

To improve buildings energy efficiency, it requires the integration of advanced HVAC and lighting systems with smart technologies and user-focused strategies. AI-driven controls, renewable energy and predictive maintenance significantly reduce energy waste and enhance comfort. The combination of adaptive shading systems, BEMS, and smart HVAC controls, with innovative materials such as bio-based insulation and phase-change materials, can rapidly decrease emissions and energy consumption. BEMS further optimize performance through real-time monitoring and automation. Together, these innovations support the shift toward more sustainable and energy-efficient buildings.

References

1. Citadini de Oliveira C, Catão Martins Vaz I, Ghisi E (2024) Retrofit strategies to improve energy efficiency in buildings: an integrative review. Energy Build 321:114624. https://doi.org/10.1016/J.ENBUILD.2024.114624
2. Hameed AI, Kakoo HS, Musa VA, Abdulla SI, Ebrahimpou A (2024) A multi-sectional analysis of building height, layout, and urban density on seasonal energy consumption: a case study. Sustain Eng Innov 6:167–176. https://doi.org/10.37868/SEI.V6I2.ID308
3. Ardente F, Beccali M, Cellura M, Mistretta M (2011) Energy and environmental benefits in public buildings as a result of retrofit actions. Renew Sustain Energy Rev 15:460–470. https://doi.org/10.1016/J.RSER.2010.09.022
4. Roberts S (2008) Altering existing buildings in the UK. Energy Policy 36:4482–4486. https://doi.org/10.1016/J.ENPOL.2008.09.023
5. Martínez B, Mendizabal V, Roncero MB, Bernat-Maso E, Gil L (2024) Towards sustainable building solutions: development of hemp shiv-based green insulation material. Constr Build Mater 414:134987. https://doi.org/10.1016/J.CONBUILDMAT.2024.134987

6. Amasyali K, El-Gohary NM (2018) A review of data-driven building energy consumption prediction studies. Renew Sustain Energy Rev 81:1192–1205. https://doi.org/10.1016/J.RSER.2017.04.095
7. Ducoulombier L, Lafhaj Z (2017) Comparative study of hygrothermal properties of five thermal insulation materials. Case Stud Therm Eng 10:628–640. https://doi.org/10.1016/J.CSITE.2017.11.005
8. Nazari S, MirzaMohammadi PK, Sajadi B, Pilehchi Ha P, Talatahari S, Sareh P (2023) Designing energy-efficient and visually-thermally comfortable shading systems for office buildings in a cooling-dominant climate. Energy Rep 10:3863–3881. https://doi.org/10.1016/J.EGYR.2023.10.062
9. Acosta I, Campano MÁ, Molina JF (2016) Window design in architecture: analysis of energy savings for lighting and visual comfort in residential spaces. Appl Energy 168:493–506. https://doi.org/10.1016/J.APENERGY.2016.02.005
10. Ahmad RM, El-Sayed Z, Taha D, Shokry H, Mahmoud H (2021) An approach to select an energy-efficient shading device for the south-oriented façades in heritage buildings in Alexandria, Egypt. Energy Rep 7:133–137. https://doi.org/10.1016/J.EGYR.2021.06.024
11. Wu S, Zhu P, Song J, Sun H, Cheng Z, Lin B (2024) Design and application optimization of static and dynamic shading technologies in multi-climate based on parameter simulation. Renew Energy 237:121846. https://doi.org/10.1016/J.RENENE.2024.121846
12. Harkouss F, Fardoun F, Biwole PH (2018) Passive design optimization of low energy buildings in different climates. Energy 165:591–613. https://doi.org/10.1016/J.ENERGY.2018.09.019
13. Es-sakali N, Zoubir Z, Idrissi Kaitouni S, Mghazli MO, Cherkaoui M, Pfafferott J (2024) Advanced predictive maintenance and fault diagnosis strategy for enhanced HVAC efficiency in buildings. Appl Therm Eng 254:123910. https://doi.org/10.1016/J.APPLTHERMALENG.2024.123910
14. Afroz Z, Shafiullah GM, Urmee T, Higgins G (2018) Modeling techniques used in building HVAC control systems: a review. Renew Sustain Energy Rev 83:64–84. https://doi.org/10.1016/J.RSER.2017.10.044
15. Aziz A, Shah SAH, Bahaidarah HMS, Zamir T, Aziz T (2025) Advanced neural network modeling with Levenberg–Marquardt algorithm for optimizing tri-hybrid nanofluid dynamics in solar HVAC systems. Case Stud Therm Eng 65:105609. https://doi.org/10.1016/J.CSITE.2024.105609
16. Nguyen AT, Pham DH, Oo BL, Santamouris M, Ahn Y, Lim BTH (2024) Modelling building HVAC control strategies using a deep reinforcement learning approach. Energy Build 310:114065. https://doi.org/10.1016/J.ENBUILD.2024.114065
17. Gala D, Khetan S, Mehendale N (2024) Assessing opportunities for enhanced lighting energy conservation via occupancy and daylight monitoring. Meas Energy 3:100015. https://doi.org/10.1016/J.MEAENE.2024.100015
18. Budhiyanto A, Chiou YS (2024) Visual comfort and energy savings in classrooms using surveillance camera derived HDR images for lighting and daylighting control system. J Build Eng 86:108841. https://doi.org/10.1016/J.JOBE.2024.108841
19. Wang Y, Yang W, Wang Q (2022) Multi-objective parametric optimization of the composite external shading for the classroom based on lighting, energy consumption, and visual comfort. Energy Build 275:112441. https://doi.org/10.1016/J.ENBUILD.2022.112441
20. Scorpio M, Ciampi G, Gentile N, Sibilio S (2022) Effectiveness of low-cost non-invasive solutions for daylight and electric lighting integration to improve energy efficiency in historical buildings. Energy Build 270:112281. https://doi.org/10.1016/J.ENBUILD.2022.112281
21. Levermore GJ (2013) Building energy management systems: applications to low-energy HVAC and natural ventilation control, 2nd edn, pp 1–519. https://doi.org/10.4324/9780203477342/BUILDING-ENERGY-MANAGEMENT-SYSTEMS-GEOFF-LEVERMORE/ACCESSIBILITY-INFORMATION
22. Karki V, Mostafa R, Gopalakrishnan B, Johnson DR (2022) Determination of effectiveness of energy management system in buildings. Energy Eng 120:561–586. https://doi.org/10.32604/EE.2023.025218

23. Fellah M, Ouhaibi S, Belouaggadia N, Mansouri K, Naji H (2024) Thermal insulation and energy performance's assessment of a mycelium-based composite wall for sustainable buildings. Case Stud Constr Mater 20:e02786. https://doi.org/10.1016/J.CSCM.2023.E02786
24. Zhang Y, Zhou C, Liu M, Li X, Liu T, Liu Z (2024) Thermal insulation performance of buildings with phase-change energy-storage wall structures. J Clean Prod 438:140749. https://doi.org/10.1016/J.JCLEPRO.2024.140749
25. Cetiner I, Shea AD (2018) Wood waste as an alternative thermal insulation for buildings. Energy Build 168:374–384. https://doi.org/10.1016/J.ENBUILD.2018.03.019
26. Schiavoni S, D'Alessandro F, Bianchi F, Asdrubali F (2016) Insulation materials for the building sector: a review and comparative analysis. Renew Sustain Energy Rev 62:988–1011. https://doi.org/10.1016/J.RSER.2016.05.045
27. Fellah M, Ouhaibi S, Belouaggadia N, Mansouri K, Sebaibi N (2024) Enhancing building energy efficiency across diverse climatic zones in Morocco through optimized eco-friendly insulation thickness with air gap. Energy Rep 12:1551–1565. https://doi.org/10.1016/J.EGYR.2024.07.040
28. Duraković B, Peštalić N, Habibović A (2024) Incorporation of phase change materials in straw-based composites: thermal characterization. In: Advanced technologies, systems, and applications IX. Springer, Cham, pp 164–171. https://doi.org/10.1007/978-3-031-71694-2_13
29. Durakovic B, Mujković N, Hudović-Kljuno A, Gibanica E (2025) Effect of phase change materials on the thermal behavior of biomass-based composites. Herit Sustain Dev 7:155–166. https://doi.org/10.37868/HSD.V7I1.1107
30. Mrajji O, Ouhaibi S, Wazna MEL, Bouari AEL, Belouaggadia N, Ezzine M, Lbibb R, Cherkaoui O (2022) Experimental and numerical investigation of eco-friendly materials for building envelope. Adv Build Energy Res 16:347–370. https://doi.org/10.1080/17512549.2021.1881615
31. Dlimi M, Iken O, Agounoun R, Zoubir A, Kadiri I, Sbai K (2019) Energy performance and thickness optimization of hemp wool insulation and air cavity layers integrated in Moroccan building walls'. Sustain Prod Consum 20:273–288. https://doi.org/10.1016/J.SPC.2019.07.008
32. Siasiev A, Dudnikov S, Nikolaienko D, Hubal H, Bondarchuk O (2025) Evaluation of the effectiveness of decision-making algorithms in the management of renewable energy sources: a case study of Ukraine. Sustain Eng Innov 7:87–96. https://doi.org/10.37868/SEI.V7I1.ID415
33. Ranjan Satpathy P, Ramachandaramurthy VK, Roslan MF, Motahhir S (2024) An adaptive architecture for strategic enhancement of energy yield in shading sensitive building-applied photovoltaic systems under real-time environments. Energy Build 324:114877. https://doi.org/10.1016/J.ENBUILD.2024.114877
34. Ashraf N, Abdin AR (2024) Biomimetic design synthesis and digital optimization of building shading skin: a novel conceptual framework for enhanced energy efficiency. Energy Build 323:114824. https://doi.org/10.1016/J.ENBUILD.2024.114824
35. Carletti C, Sciurpi F, Pierangioli L, Asdrubali F, Pisello AL, Bianchi F, Sambuco S, Guattari C (2016) Thermal and lighting effects of an external venetian blind: experimental analysis in a full scale test room. Build Environ 106:45–56. https://doi.org/10.1016/J.BUILDENV.2016.06.017
36. Park JH, Yun BY, Chang SJ, Wi S, Jeon J, Kim S (2020) Impact of a passive retrofit shading system on educational building to improve thermal comfort and energy consumption. Energy Build 216:109930. https://doi.org/10.1016/J.ENBUILD.2020.109930
37. Ben Romdhane S, Amamou A, Ben Khalifa R, Saïd NM, Younsi Z, Jemni A (2020) A review on thermal energy storage using phase change materials in passive building applications. J Build Eng 32:101563. https://doi.org/10.1016/J.JOBE.2020.101563
38. Durakovic B, Torlak M (2017) Experimental and numerical study of a PCM window model as a thermal energy storage unit. Int J Low-Carbon Technol 12:272–280. https://doi.org/10.1093/ijlct/ctw024
39. Duraković B, Mešetović S (2019) Thermal performances of glazed energy storage systems with various storage materials: an experimental study. Sustain Cities Soc 45:422–430. https://doi.org/10.1016/j.scs.2018.12.003

40. HVAC ventilation strategies: the contribution for thermal comfort, energy efficiency, and indoor air quality. J Green Build. https://meridian.allenpress.com/jgb/article/2/2/131/200076/HVAC-Ventilation-Strategies-The-Contribution-for. Accessed 2024/12/08
41. Khan O, Parvez M, Seraj M, Yahya Z, Devarajan Y, Nagappan B (2024) Optimising building heat load prediction using advanced control strategies and artificial intelligence for HVAC system. Therm Sci Eng Prog 49:102484. https://doi.org/10.1016/J.TSEP.2024.102484
42. Rawat N, Kumar P (2023) Interventions for improving indoor and outdoor air quality in and around schools. Sci Total Environ 858:159813. https://doi.org/10.1016/J.SCITOTENV.2022.159813
43. Dimitroulopoulou S, Dudzińska MR, Gunnarsen L, Hägerhed L, Maula H, Singh R, Toyinbo O, Haverinen-Shaughnessy U (2023) Indoor air quality guidelines from across the world: an appraisal considering energy saving, health, productivity, and comfort. Environ Int 178:108127. https://doi.org/10.1016/J.ENVINT.2023.108127
44. Gibbons L, Javed S (2023) Cost-optimal analysis of energy efficient HVAC solution-sets for low-energy apartment buildings. Energy Build 295:113332. https://doi.org/10.1016/J.ENBUILD.2023.113332
45. Shamoon A, Haleem A, Bahl S, Javaid M, Bala Garg S (2022) Role of energy technologies in response to climate change. Mater Today Proc 62:63–69. https://doi.org/10.1016/J.MATPR.2022.02.025
46. Shoeibi S, Kargarsharifabad H, Sadi M, Arabkoohsar A, Mirjalily SAA (2022) A review on using thermoelectric cooling, heating, and electricity generators in solar energy applications. Sustain Energy Technol Assess 52:102105. https://doi.org/10.1016/J.SETA.2022.102105
47. Han Y, Li J, Lou X, Fan C, Geng Z (2022) Energy saving of buildings for reducing carbon dioxide emissions using novel dendrite net integrated adaptive mean square gradient. Appl Energy 309:118409. https://doi.org/10.1016/J.APENERGY.2021.118409
48. Chen J, Zhang L, Li Y, Shi Y, Gao X, Hu Y (2022) A review of computing-based automated fault detection and diagnosis of heating, ventilation and air conditioning systems. Renew Sustain Energy Rev 161:112395. https://doi.org/10.1016/J.RSER.2022.112395
49. Chen Z, O'Neill Z, Wen J, Pradhan O, Yang T, Lu X, Lin G, Miyata S, Lee S, Shen C, Chiosa R, Piscitelli MS, Capozzoli A, Hengel F, Kührer A, Pritoni M, Liu W, Clauß J, Chen Y, Herr T (2023) A review of data-driven fault detection and diagnostics for building HVAC systems. Appl Energy 339:121030. https://doi.org/10.1016/J.APENERGY.2023.121030
50. Wang D, Zheng W, Wang Z, Wu Z, Shen B, Tian S (2024) Quantifying the potential of load flexibility for building HVAC system using model predictive control strategy. Energy Build 323:114819. https://doi.org/10.1016/J.ENBUILD.2024.114819
51. Le Dréau J, Heiselberg P (2016) Energy flexibility of residential buildings using short term heat storage in the thermal mass. Energy 111:991–1002. https://doi.org/10.1016/J.ENERGY.2016.05.076
52. Gao C, Wang D (2023) Comparative study of model-based and model-free reinforcement learning control performance in HVAC systems. J Build Eng 74:106852. https://doi.org/10.1016/J.JOBE.2023.106852
53. Deng X, Zhang Y, Qi H (2022) Towards optimal HVAC control in non-stationary building environments combining active change detection and deep reinforcement learning. Build Environ 211:108680. https://doi.org/10.1016/J.BUILDENV.2021.108680
54. Shaikh PH, Nor NBM, Nallagownden P, Elamvazuthi I, Ibrahim T (2014) A review on optimized control systems for building energy and comfort management of smart sustainable buildings. Renew Sustain Energy Rev 34:409–429. https://doi.org/10.1016/J.RSER.2014.03.027
55. CBECS 2012: trends in lighting in commercial buildings. https://www.eia.gov/consumption/commercial/reports/2012/lighting/?utm_source=chatgpt.com. Accessed 2025/04/11
56. Uyan E, Atlar M, Ölçer AI (2023) Improving the energy efficiency of lighting systems for a marine equipment manufacturing plant through retrofitting, daylighting, and behaviour change. J Clean Prod 413:137216. https://doi.org/10.1016/J.JCLEPRO.2023.137216
57. Doulos LT, Kontadakis A, Madias EN, Sinou M, Tsangrassoulis A (2019) Minimizing energy consumption for artificial lighting in a typical classroom of a Hellenic public school aiming for

near zero energy building using LED DC luminaires and daylight harvesting systems. Energy Build 194:201–217. https://doi.org/10.1016/J.ENBUILD.2019.04.033
58. Mahmoud FA, Hanoon MF (2025) Analysis of reducing unused energy costs by leveraging ABC and TABC methods for sustainable operations. Herit Sustain Dev 7:211–222. https://doi.org/10.37868/HSD.V7I1.1089
59. Xiong J, Tzempelikos A, Bilionis I, Karava P (2019) A personalized daylighting control approach to dynamically optimize visual satisfaction and lighting energy use. Energy Build 193:111–126. https://doi.org/10.1016/J.ENBUILD.2019.03.046
60. Gentile N, Laike T, Dubois MC (2016) Lighting control systems in individual offices rooms at high latitude: measurements of electricity savings and occupants' satisfaction. Sol Energy 127:113–123. https://doi.org/10.1016/J.SOLENER.2015.12.053
61. AlFaris F, Juaidi A, Manzano-Agugliaro F (2016) Improvement of efficiency through an energy management program as a sustainable practice in schools. J Clean Prod 135:794–805. https://doi.org/10.1016/J.JCLEPRO.2016.06.172
62. Nardelli A, Deuschle E, de Azevedo LD, Pessoa JLN, Ghisi E (2017) Assessment of light emitting diodes technology for general lighting: a critical review. Renew Sustain Energy Rev 75:368–379. https://doi.org/10.1016/J.RSER.2016.11.002
63. Winterbottom M, Wilkins A (2009) Lighting and discomfort in the classroom. J Environ Psychol 29:63–75. https://doi.org/10.1016/J.JENVP.2008.11.007
64. Anand P, Cheong D, Sekhar C, Santamouris M, Kondepudi S (2019) Energy saving estimation for plug and lighting load using occupancy analysis. Renew Energy 143:1143–1161. https://doi.org/10.1016/J.RENENE.2019.05.089
65. Belany P, Hrabovsky P, Kolkova Z (2021) Combination of lighting retrofit and life cycle cost analysis for energy efficiency improvement in buildings. Energy Rep 7:2470–2483. https://doi.org/10.1016/J.EGYR.2021.04.044
66. de Souza DF, da Silva PPF, Fontenele LFA, Barbosa GD, de Oliveira Jesus M (2019) Efficiency, quality, and environmental impacts: a comparative study of residential artificial lighting. Energy Rep 5:409–424. https://doi.org/10.1016/J.EGYR.2019.03.009
67. Gentile N (2022) Improving lighting energy efficiency through user response. Energy Build 263:112022. https://doi.org/10.1016/J.ENBUILD.2022.112022
68. BEMS: understanding building energy management systems. BMS Controls and Energy. https://bmscontrols.co.uk/blog/bems-understanding-building-energy-management-systems/?utm_source=chatgpt.com. Accessed 2025/04/11

Chapter 4
Building Management Systems

Abbreviations

BMS	Building Management system
CCTV	Closed-Circuit Television
EU	European Union
GHG	Greenhouse gas
HVAC	Heating, Ventilation and Air conditioning
MOO	Multi-Objective Optimization
RL	Reinforcement learning
ZEB	Zero Emission building

4.1 Introduction

The building sector has long been a significant contributor to global energy consumption and carbon emissions, positioning it as a key area for addressing climate challenges [1, 2]. On the other hand, fossil fuels have been major enabler of industrial development which still dominates in the global energy mix, even though ongoing efforts toward energy efficiency and renewables [3–5]. Therefore, improving building energy efficiency is considered as an effective strategy for reducing greenhouse gas emissions but it requires the integration of smart systems, improved building envelope and behavioral changes [6, 7].

In this context, the integration of smart systems such as Building Management Systems (BMS) and advanced energy management models that utilize predictive algorithms and optimization techniques have been developed [8, 9]. The purpose of these smart systems is to monitor energy usage in real time and optimize energy efficiency by reducing energy losses [10]. The BMS controls and monitors various devices and systems integrated into the building operations, including advanced

B. Duraković, *Advancing Zero Energy Buildings*, Green Energy and Technology,
https://doi.org/10.1007/978-3-032-16601-2_4

Building Automation Systems, enabling unified control of Heating, Ventilation and Air conditioning (HVAC), lighting, access systems, fire safety, Closed-Circuit Television (CCTV), elevators, and energy monitoring. The enhancement of these technologies with AI and IoT, helps optimize operations, improve energy efficiency, and ensure occupant comfort through real-time analyses and predictive maintenance, managing the growing need for sustainable energy solutions in the building sector [11–13]. Therefore, BMS connects and makes different building systems work together and provides an effective solution to reducing energy consumption and CO_2 emissions [14, 15].

BMS provides an interface for building operators that can see and control everything [16]. The systems can monitor and adjust how the building is operating, even from a remote location. It can send alerts in case of an error, maintain history of how the system has been working, control who can use the system, and track the actions that the users obtain. All these features help building staff save time since physical presence is not required for adjusting or repairing the equipment. Most of these tasks, whether they include configuration or checking devices, can be performed remotely, hence it reduces the need for large team [17]. Optimization models are what makes a BMS smart and capable of making intelligent decisions. The models use control algorithms that are integrated into the BMS platform and need information such as how much energy is being used and predict renewable energy availability [18]. With this information it sends commands to the equipment in the building to optimize performance and energy use. For this to be effective, BMS needs models that will quickly represent how the building heating, cooling, and electrical systems behave [19, 20]. The crucial point is that these models must be fast enough to be included in the optimization process so that the BMS can make required adjustments to improve the system performance [21]. On the contrary limited human resources can make serious errors to occur when the BMS is not working properly, mainly because of limited staff members. This causes the occurred errors to go unnoticed for longer periods of time [22].

4.2 The Role of Building Management Systems in ZEBs

Buildings around the world account for nearly 36% of total energy consumption and 26% of energy is related to Greenhouse Gases (GHG) [23]. In the European Union (EU), the buildings consume around 40% of all energy and are responsible for around one-third of the related greenhouse gas emissions. This became a key focus for energy and climate policies in big economies. In the last decades progress has been seen in reducing carbon emissions through the adoption of strategies addressing the entire life cycle of buildings. These included all stages of building life cycle from material selection to operational efficiency, supporting climate neutrality goals [24]. Figure 4.1 shows the percentage of energy consumption and greenhouse gas emissions assigned to buildings globally and in the EU, while emphasizing the sector's significant environmental effect.

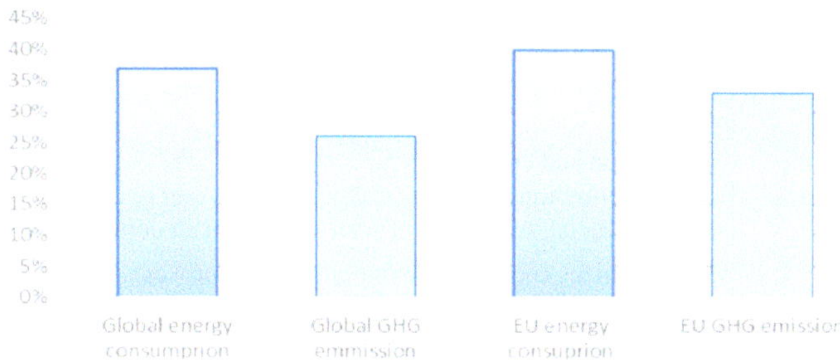

Fig. 4.1 Building energy consumption and greenhouse gas emissions in the EU[1]

Referring to Fig. 4.1 it is observed that the buildings play an important role in international and local energy consumption and GHG emissions, emphasizing the building sector's critical significance in managing energy and climate challenges. This implies a need for innovations in energy-efficiency designs, the adoption of renewable energy technologies and regulations promoting sustainable construction practices.

The topic of this chapter covers the importance of the zero-energy concept as a solution to reduce energy use and GHG emissions. Zero Energy Buildings (ZEBs) are designed to be nearly or net zero energy buildings able to generate as much energy as they consume [25]. Before including renewable energy technologies, it is important to improve building energy efficiency [26]. Achieving ZEB features requires theoretical knowledge and practical implementation as well as energy system monitoring like main energy conversion elements, assist comparing energy use across different contexts [27]. Another important aspect is evaluating energy use over time, which is crucial for long-term efficiency. Table 4.1 shows key aspects of energy consumption in the building sector, the role of BMS, the significant carbon emissions, BMS's ability to reduce energy use, the challenges of relying on fossil fuel, and the need for fast and accurate models.

Referring to Table 4.1 it is observed that the contribution of BMS is optimizing energy usage, improving efficiency, and allowing remote monitoring and control. Technological innovations are related to the use of smart optimization models to assist BMS to modify performance based on real-time information. Experiments related to energy use in shorter time periods provide more precise results in detecting energy imbalances that are not detected in yearly evaluations [32]. BMS plays an important role in advancing ZEBs by optimizing energy usage, incorporating renewable energy sources, enabling demand response and energy storage. BMS enhances the management of HVAC and lighting systems, occupant comfort, and building operations through real time monitoring and analytics [33, 34].

[1] Created by the author based on [23, 24].

Table 4.1 Energy efficiency and technological innovations aspects in BMS

Aspects	Results	References
Energy consumption and carbon emissions	– The building sector accounts for about 40% of global energy consumption and emission – Reducing emissions through energy efficiency and renewable energy is crucial, regardless of fossil fuels remain dominant	[1, 4, 5] [6, 7] [28, 29]
Building management systems	– BMS optimizes energy usage in real-time, improves efficiency, reduces consumption, allows remote management and error alerts	[9, 11, 30] [12, 31] [16, 17]
Technological innovations	– BMS uses smart optimization models to adjust performance, based on energy use and renewable energy availability	[18–20]
Optimization models and efficiency	– Fast, accurate models are important for optimizing building systems – Limited human resources can delay error detection	[21, 22]

4.3 Benefits of a Building Management System

BMS enhances building intelligence and minimizes operating costs through the integration and control of renewable energy generation, energy storage and HVAC systems [35, 36]. In addition, BMS improves grid stability by adjusting energy demand and consumption enabling buildings to interact effectively with the grid [37]. Integration of BMS in existing buildings through retrofitting can achieve energy savings up to 80% while adjusting operations to real-time energy needs, climate conditions, and grid connections [38]. Figure 4.2 shows how BMS reduce energy consumption and CO_2 emissions compared to buildings without BMS.

Referring to Fig. 4.2, it is observed that significant energy saving, and carbon reduction can be achieved with BMS implementation. Particularly, 30% of reduction in energy consumption when BMS is implemented and about 35% reduction in CO_2 emissions. These insights highlight the crucial role of BMS in optimizing building operations, minimizing energy waste, and reducing the environmental impact. Equation (4.1) quantifies the total impact of BMS by combining

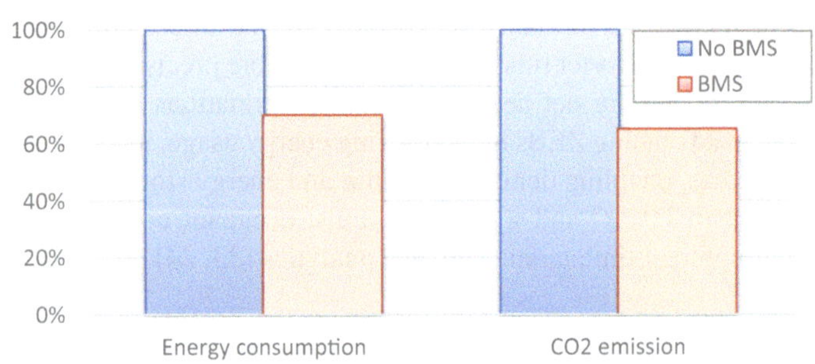

Fig. 4.2 Impact of building management systems on energy consumptions and CO_2 emissions (created by the author based on [39])

$$BMS_{Impact} = E_{Savings} + C_{Reduction} + OE_{Improvement} \tag{4.1}$$

where, $E_{Savings}$ is energy savings achieved through BMS optimization (e.g., real-time monitoring, resource management), $C_{Reduction}$ is carbon reduction due to optimized energy use, and $OE_{Improvement}$ is operational efficiency improvements such as reduced maintenance cost, remote monitoring, and task optimization.

Advanced occupancy-based technologies based on smart sensors and powered with AI additionally boost operational performance by adjusting lighting and HVAC in relation to real-time occupancy schemes [40]. The dynamic energy assessments enables a more precise and sustainable strategy to review energy efficiency by including hourly examinations. This affects buildings' maintaining balance energy supply and demand and further remove on-site carbon emissions by providing stricter decarbonization plans [41]. However, the incorporation of machine learning with the Last Planner System gives a predictive view that allows project teams to predict schedule variations, address risks responsively, and sustain timelines, enhancing decision-making and developing endurance in complex projects [42]. The aforementioned benefits indicate rapid developments in building management systems. The incorporation of environmental, economic and operational profits provides a basis for a more intelligent, efficient, and sustainable built environment.

4.4 Trends in Building Management Systems BMS in ZEBs

Fighting the climate changes was further reinforced in 2015 when 195 countries signed a historic document in Paris by reducing greenhouse gas emissions and restricting global temperature rise. The agreement focused on the need for low-emission technologies, and 20 countries, including United States, China and several European nations, pledged to increase their spending on clean energy research over the next five years [43]. The building sector, which absorbs more than 30% of global energy consumption, became a highlight of these initiatives [44]. One strategy, ZEBs, intends to decrease fossil fuel use and confront the Urban Heat Island effect. ZEBs use renewable energy and smart systems to enhance performance, energy generation, and storage, supporting sustainability and smart-grid technology [45, 46]. Renewable energy systems contribute to reducing greenhouse gas emissions and control heat from human activities. Distributed Energy Resources in buildings produce or retain renewable energy, but a challenge for ZEBs is balancing energy demand and supply. Advanced control systems, like Demand Side Management, maximize energy use in real time [47].

The BMS is often connected with smart micro-grids to manage energy use and respond to real time situations. Developed control methods, like AI, model predictive control and multi-agent systems, present flexible solutions, permitting buildings to control energy efficiently while managing occupant comfort [39]. Research into computational optimization methods is being introduced within the architecture, engineering, and construction (AEC) industry, with a focal point on bettering building performance optimization [48]. The aim of optimization is to reach the ideal approach that will maximize or minimize these functions within the limitations [49]. Traditionally, single-objective optimization, such as reducing cost or carbon impact, has been the primary focus [50, 51]. However, the way building systems are becoming more complex, there is increasing shift at multi-objective optimization (MOO). MOO is helpful for managing opposing goals, like balancing energy use with comfort, cost or emissions. Rather than discovering an ideal solution, MOO is searching for a range of Pareto optimal answers [52]. This introduction is integrated in many building design areas, involving envelope design, HVAC systems, renewable energy integration, and building management systems [53].

Latest trends introduce the increased use of evolutionary algorithms, such as Genetic Algorithms and Particle Swarm Optimization, thanks to their performance in resolving MOO problems. For Zero Energy Buildings (ZEBs), improving performance includes advanced analytical techniques and intelligent control methods [54]. The focus on MOO indicates the difficulty of modern building systems, where several, mostly opposing goals must be harmonized. This strategy supports sustainable building methods by upgrading energy efficiency, decreasing emissions, and increasing occupant comfort, all in economical way [55].

4.5 Challenges in Implementing BMS in ZEBs

Modeling energy systems for buildings is complex, as it involves many factors related to heating, cooling, other engineering systems, occupant behavior, building materials, and environmental factors. Advanced occupant comfort is often associated with increased energy use and material consumption, which caused the increase in costs and pollution. Although these developments in energy systems are designed for efficiency, it is important to note that they do not always result in reduced environmental impact when their full life cycle is considered.

Integrating renewable energy sources and Economic Model Predictive Control increases the complexity of the system. The purpose of these systems is to balance energy demand, optimize performance and resolve uncertainties in smart buildings [56, 57]. This way energy use and occupant comfort were more successfully balanced than with conventional controls [58].

Challenges associated with implementations BMS include maintaining operational reliability under dynamic real-world conditions. This includes climate variability and shifting occupancy conditions, compatibility with existing infrastructure and the integration of adaptive energy management strategies with varied building systems.

Majority of energy models today are based on static data and assumptions limiting their ability to represent real-world performance. Adapted systems representing dynamic real-world conditions such as weather and occupancy changes have been introduced. One of them is Reinforcement Learning (RL), but they are complex to implement and require extensive training and testing. Therefore, an effective *energy management* system should have the right balance between energy optimization and indoor comfort [59]. Traditional PID-based controllers underperform since they do not adjust to these changing conditions and therefore require manual adjustments. Therefore, PID controllers can be unreliable and lead to fluctuations and instability in more complex dynamic systems [60].

RL-based systems face other challenges like the need for computational power, training times, and huge amounts of data for model training [61, 62]. Other challenges of RL integration in existing BMS are *compatibility* with existing infrastructure and the need for long testing to achieve reliable performance [63]. The lack of cost-benefit analyses and long-term validation has slowed the adoption of RL in BMS [64]. Regardless of these obstacles, RL based systems have notable potential for reducing energy consumption. Advanced optimization methodologies such as Multi-Objective Particle Swarm Optimization have limited adoption due to the need for extensive computational power as well [65].

To address these challenges, multifaceted approach is required including demand for better training methods, adaptive neural network designs, and improved reward functions [66]. To overcome the complexities of practical deployment of RL-based energy management systems, ongoing research and development are important to secure reliable long-term performance and provide a detailed cost-benefit analysis [67]. Table 4.2 provides a summary of energy management systems challenges including limitations of traditional controls and RL's computational demands as well as directions for improving RL adaptability for broader adoption.

Table 4.2 Challenges and future directions in BMS

Challenges	Results	References
Current challenges	– Requirement of understanding energy systems – Enhancement of comfort increases cost, pollutants and energy use – Static models lack real-time application	[56, 68, 69]
Control methods	– Traditional methods (e.g., PID controllers) are not adaptive to changing conditions – Requires manual adjustment and may lead to inefficiency and instability	[60, 70]
Challenges of RL	– RL requires higher computational power, large data sets and extensive testing – Integration with existing systems can be very challenging – Lacks comprehensive cost-benefit analysis and long-term validation	[61–64, 69]
Future directions	– Focus on improving the RL's adaptability, efficiency, and integration with existing systems – Development of advanced training methods, neutral networks, and cost-benefit analyses for RL systems – Address computational and testing challenges to enable large-scale adoption	[66, 67]

Referring to Table 4.2 it is observed that one of the major challenges in BMS is inefficiency of traditional control methods, the extensive RL computational demands, integration with existing BMS restricting its widespread adoption. Therefore, to support wider adoption of LR it is necessary to improve adaptability training methods to reduce computational demands.

4.6 Conclusion

Building Management Systems play a crucial role in transforming buildings into more energy-efficient and more sustainable structures. It helps to reduce energy consumption through optimization of resource use and minimization of carbon emissions. BMS controls and monitors various devices and systems that are integrated into the building operations such as advanced BAS, HVAC, lighting, access systems, fire safety, CCTV, elevators, and energy monitoring. The BMS provides real-time monitoring of building operators and provides important insights within system performance, empowering better decision making and cost savings. In addition, operational efficiency is increased since the BMS optimizes tasks and allows remote management which reduces the need for large maintenance teams. Overall, BMS are essential to ZEB development since it plays crucial role in managing energy production, storage, and demand responses.

References

1. Brandi S, Gallo A, Capozzoli A (2022) A predictive and adaptive control strategy to optimize the management of integrated energy systems in buildings. Energy Rep 8:1550–1567. https://doi.org/10.1016/j.egyr.2021.12.058
2. Fellah M, Ouhaibi S, Belouaggadia N, Mansouri K, Sebaibi N (2024) Enhancing building energy efficiency across diverse climatic zones in marocco through optimized eco-friendly insulation thickness with air gap. Energy Rep 12:1551–1565. https://doi.org/10.1016/J.EGYR.2024.07.040
3. Bilous I, Biriukov D, Karpenko D, Eutukhova T, Novoseltsev O, Voloshchuk V (2024) Reinforcement learning model for energy system management to ensure energy efficiency and comfort in buildings. Energy Eng 121:3617–3634. https://doi.org/10.32604/EE.2024.051684
4. Dehghan Shabani Z (2024) Renewable energy and CO_2 emissions: does human capital matter? Energy Rep 11:3474–3491. https://doi.org/10.1016/j.egyr.2024.03.021
5. González-Torres M, Pérez-Lombard L, Coronel JF, Maestre IR, Yan D (2022) A review on buildings energy information: trends, end-uses, fuels and drivers. https://doi.org/10.1016/j.egyr.2021.11.280
6. Andolfi L, Lima Baima R, Burcheri LM, Pavić I, Fridgen G (2025) Sociotechnical design of building energy management systems in the public sector: five design principles. Appl Energy 377. https://doi.org/10.1016/j.apenergy.2024.124628
7. Ali MI, Islam MM, Ceh B (2024) Green versus grey: impact of renewable and non-renewable energy usage on Canada's growth trajectory in the context of internal and external forces. Sustain Futures 8. https://doi.org/10.1016/j.sftr.2024.100258

8. Jing T, Zhao Y (2024) Optimizing energy consumption in smart buildings: a model for efficient energy management and renewable integration. Energy Build 323:114754. https://doi.org/10.1016/J.ENBUILD.2024.114754
9. Costa R, Silva R, Faia R, Gomes L, Faria P, Vale Z (2024) Empowering energy management in smart buildings: a comprehensive study on distributed energy storage systems for sustainable consumption. Energy Build 324. https://doi.org/10.1016/j.enbuild.2024.114953
10. Dange KM (2023) A comprehensive review on IoT based smart cities. Sustain Eng Innov 5:48–57. https://doi.org/10.37868/SEI.V5I1.ID188
11. Bianco G, Delfino F, Ferro G, Robba M, Rossi M (2023) A hierarchical building management system for temperature's optimal control and electric vehicles' integration. Energy Convers Manag X 17. https://doi.org/10.1016/j.ecmx.2022.100339
12. Annebicque D, Robert B, Henry JF, Randrianalisoa J, Popa C (2016) A multidisciplinary approach to improve energetic performance in smart buildings. IFAC-PapersOnLine 313–317. https://doi.org/10.1016/j.ifacol.2016.10.710
13. Saleem M, Khalid H, Kamarudin S, Almaslukh FM, Subramaniam RT, Wijenayake A (2025) Integration of new age technology with smart cities initiatives and its role in the achievement of sustainable development goals. Herit Sustain Dev 7:589–610. https://doi.org/10.37868/HSD.V7I1.1097
14. Pallonetto F, De Rosa M, D'Ettorre F, Finn DP (2020) On the assessment and control optimisation of demand response programs in residential buildings. Renew Sustain Energy Rev 127:109861. https://doi.org/10.1016/J.RSER.2020.109861
15. Aguilar J, Garces-Jimenez A, R-Moreno MD, García R (2021) A systematic literature review on the use of artificial intelligence in energy self-management in smart buildings. https://doi.org/10.1016/j.rser.2021.111530
16. Yang T, Clements-Croome D, Marson M (2024) Building energy management systems. In: Encyclopedia of sustainable technologies, pp 727–749. https://doi.org/10.1016/B978-0-323-90386-8.00025-5
17. Habib M, Bollin E, Wang Q (2023) Edge-based solution for battery energy management system: investigating the integration capability into the building automation system. J Energy Storage 72. https://doi.org/10.1016/j.est.2023.108479
18. Zheng Z, Marie S, Farazdaghi E, Yahia E, Makhoul K, Lagarde T, El Meouche R, Ababsa F (2024) Mastering building management systems data points tagging with minimal examples: unveiling the power of large language models. Energy Build 115173. https://doi.org/10.1016/J.ENBUILD.2024.115173
19. Graveto V, Cruz T, Simöes P (2022) Security of building automation and control systems: survey and future research directions. Comput Secur 112. https://doi.org/10.1016/j.cose.2021.102527
20. Gao T, Marié S, Béguery P, Thebault S, Lecoeuche S (2024) Integrated building fault detection and diagnosis using data modeling and Bayesian networks. Energy Build 306:113889. https://doi.org/10.1016/J.ENBUILD.2024.113889
21. Pilanawithana NM, Feng Y, London K, Zhang P (2022) Developing resilience for safety management systems in building repair and maintenance: a conceptual model. Saf Sci 152:105768. https://doi.org/10.1016/J.SSCI.2022.105768
22. Kučera A, Glos P, Pitner T (2013) Fault detection in building management system networks. IFAC Proc Vol (IFAC-PapersOnLine) 416–421. https://doi.org/10.3182/20130925-3-CZ-3023.00027
23. Maduta C, D'Agostino D, Tsemekidi-Tzeiranaki S, Castellazzi L (2024) From nearly zero-energy buildings (NZEBs) to zero-emission buildings (ZEBs): current status and future perspectives. Energy Build 115133. https://doi.org/10.1016/j.enbuild.2024.115133
24. Alaux N, Marton C, Steinmann J, Maierhofer D, Mastrucci A, Petrou D, Potrč Obrecht T, Ramon D, Le Den X, Allacker K, Passer A, Röck M (2024) Whole-life greenhouse gas emission reduction and removal strategies for buildings: impacts and diffusion potentials across EU Member States. J Environ Manag. https://doi.org/10.1016/j.jenvman.2024.122915

25. D'Agostino D, Mazzarella L (2019) What is a nearly zero energy building? Overview, implementation and comparison of definitions. J Build Eng 21:200–212. https://doi.org/10.1016/J.JOBE.2018.10.019
26. Citadini de Oliveira C, Catão Martins Vaz I, Ghisi E (2024) Retrofit strategies to improve energy efficiency in buildings: an integrative review. Energy Build 321:114624. https://doi.org/10.1016/J.ENBUILD.2024.114624
27. Junker RG, Azar AG, Lopes RA, Lindberg KB, Reynders G, Relan R, Madsen H (2018) Characterizing the energy flexibility of buildings and districts. Appl Energy 225:175–182. https://doi.org/10.1016/J.APENERGY.2018.05.037
28. Tauseef Hassan S, Wang P, Khan I, Zhu B (2023) The impact of economic complexity, technology advancements, and nuclear energy consumption on the ecological footprint of the USA: towards circular economy initiatives. Gondwana Res 113:237–246. https://doi.org/10.1016/J.GR.2022.11.001
29. Xiang X, Zhou N, Ma M, Feng W, Yan R (2023) Global transition of operational carbon in residential buildings since the millennium. Adv Appl Energy 11. https://doi.org/10.1016/j.adapen.2023.100145
30. Nnaji C, Karakhan AA (2020) Technologies for safety and health management in construction: current use, implementation benefits and limitations, and adoption barriers. J Build Eng 29:101212. https://doi.org/10.1016/J.JOBE.2020.101212
31. Seraj M, Parvez M, Khan O, Yahya Z (2024) Optimizing smart building energy management systems through industry 4.0: a response surface methodology approach. Green Technol Sustain 2:100079. https://doi.org/10.1016/J.GRETS.2024.100079
32. Abdous M, Aslani A, Noorollahi Y, Zahedi R, Yousefi H (2024) Design and analysis of zero-energy and carbon buildings with renewable energy supply and recycled materials. Energy Build 324:114922. https://doi.org/10.1016/J.ENBUILD.2024.114922
33. Olasolo-Alonso P, López-Ochoa LM, Las-Heras-Casas J, López-González LM (2023) Energy performance of buildings directive implementation in southern European countries: a review. Energy Build 281:112751. https://doi.org/10.1016/J.ENBUILD.2022.112751
34. Ibrahim M, Harkouss F, Biwole P, Fardoun F, Ouldboukhitine S (2024) Building retrofitting towards net zero energy: a review. Energy Build 322:114707. https://doi.org/10.1016/J.ENBUILD.2024.114707
35. Liu C, Luo B, Wang W, Gao H, Wang Z, Ding H, Yu M, Peng Y (2022) Energy management and capacity optimization of photovoltaic, energy storage system, flexible building power system considering combined benefit. Energy Eng 120:541–559. https://doi.org/10.32604/EE.2023.022610
36. Durakovic B, Mujković N, Hudović-Kljuno A, Gibanica E (2025) Effect of phase change materials on the thermal behavior of biomass-based composites. Herit Sustain Dev 7:155–166. https://doi.org/10.37868/HSD.V7I1.1107
37. Li H, Hong T (2022) A semantic ontology for representing and quantifying energy flexibility of buildings. Adv Appl Energy 8:100113. https://doi.org/10.1016/J.ADAPEN.2022.100113
38. Al Dakheel J, Del Pero C, Leonforte F, Aste N, El Mankibi M (2024) Assessing the performance of smart buildings and smart retrofit interventions through key performance indicators: defining minimum performance thresholds. Energy Build 325:114988. https://doi.org/10.1016/J.ENBUILD.2024.114988
39. Gnekpe C, Tchuente D, Nyawa S, Dey PK (2024) Energy performance of building refurbishments: predictive and prescriptive AI-based machine learning approaches. J Bus Res 183:114821. https://doi.org/10.1016/J.JBUSRES.2024.114821
40. Norouziasas A, Attia S, Hamdy M (2024) Impact of space utilization and work time flexibility on energy performance of office buildings. J Build Eng 98. https://doi.org/10.1016/j.jobe.2024.111032
41. Bilardo M, Kämpf JH, Fabrizio E (2024) From zero energy to zero power buildings: a new paradigm for a sustainable transition of the building stock. Sustain Cities Soc 101. https://doi.org/10.1016/j.scs.2023.105136

42. Lagos CI, Herrera RF, Mac Cawley AF, Alarcón LF (2024) Predicting construction schedule performance with last planner system and machine learning. Autom Constr 167:105716. https://doi.org/10.1016/J.AUTCON.2024.105716
43. Zheng Z, Wang L, Hienwong N (2016) Intelligent control system integration and optimization for zero energy buildings to mitigate urban heat island. Procedia Eng 169:100–107. https://doi.org/10.1016/J.PROENG.2016.10.012
44. Chen S, Zhang G, Xia X, Setunge S, Shi L (2020) A review of internal and external influencing factors on energy efficiency design of buildings. Energy Build 216:109944. https://doi.org/10.1016/J.ENBUILD.2020.109944
45. Danza L, Belussi L, Guazzi G, Meroni I, Salamone F (2018) Durability of technologies in the keeping of ZEB's performances. Energy Procedia 148:138–145. https://doi.org/10.1016/J.EGYPRO.2018.08.041
46. Hawks MA, Cho S (2024) Review and analysis of current solutions and trends for zero energy building (ZEB) thermal systems. Renew Sustain Energy Rev 189:114028. https://doi.org/10.1016/J.RSER.2023.114028
47. Nasab MA, Hatami M, Zand M, Nasab MA, Padmanaban S (2024) Demand side management programs in smart grid through cloud computing. Renew Energy Focus 51:100639. https://doi.org/10.1016/J.REF.2024.100639
48. Aritonang PK, Wiryono SK, Faturohman T (2025) Hidden-layer configurations in reinforcement learning models for stock portfolio optimization. Intell Syst Appl 25:200467. https://doi.org/10.1016/J.ISWA.2024.200467
49. Vojdani B, Rahbar M, Fazeli M, Hakimazari M, Samuelson HW (2024) Comparative study of optimization methods for building energy consumption and daylighting performance. Energy Build 323:114753. https://doi.org/10.1016/J.ENBUILD.2024.114753
50. Hosamo H, Coelho GBA, Nordahl Rolfsen C, Kraniotis D (2024) Building performance optimization through sensitivity analysis, and economic insights using AI. Energy Build 325:114999. https://doi.org/10.1016/J.ENBUILD.2024.114999
51. Jiang R, Shankaran R, Wang S, Chao T (2022) A proportional, integral and derivative differential evolution algorithm for global optimization. Expert Syst Appl 206:117669. https://doi.org/10.1016/J.ESWA.2022.117669
52. Zhao T, Zhuang Z, Chang T, Grunewald J, Yu Y (2025) Performance-oriented design of thermal-activated building envelope based on multi-objective optimization: method and application. Energy Build 328:115198. https://doi.org/10.1016/J.ENBUILD.2024.115198
53. Sun Y, Luo Z, Li Y, Zhao T (2024) Grey-box model-based demand side management for rooftop PV and air conditioning systems in public buildings using PSO algorithm. Energy 296:131052. https://doi.org/10.1016/J.ENERGY.2024.131052
54. Hernandez-Robles IA, González-Ramírez X, Antonio Álvarez-Jaime J (2024) Effectiveness of forecasters based on neural networks for energy management in zero energy buildings. Energy Build 316:114372. https://doi.org/10.1016/J.ENBUILD.2024.114372
55. Hua Z, Zhang L, Zhang S, Yang S, Liu C (2024) Multi-objective coordinated optimization of low-carbon building energy systems based on high renewable energy penetration. J Build Eng 96:110577. https://doi.org/10.1016/J.JOBE.2024.110577
56. Fernandez MI, Go YI, Wong DML, Früh WG (2024) Review of challenges and key enablers in energy systems towards net zero target: renewables, storage, buildings, & grid technologies. Heliyon 10:e40691. https://doi.org/10.1016/J.HELIYON.2024.E40691
57. Zong Y, Böning GM, Santos RM, You S, Hu J, Han X (2017) Challenges of implementing economic model predictive control strategy for buildings interacting with smart energy systems. Appl Therm Eng 114:1476–1486. https://doi.org/10.1016/J.APPLTHERMALENG.2016.11.141
58. Liu H, Liu P, Bai C (2025) Combining long and short spatiotemporal reasoning for deep reinforcement learning. Neurocomputing 619:129165. https://doi.org/10.1016/J.NEUCOM.2024.129165
59. Bai L, Tan Z (2024) Optimizing energy efficiency, thermal comfort, and indoor air quality in HVAC systems using a robust DRL algorithm. J Build Eng 98:111493. https://doi.org/10.1016/J.JOBE.2024.111493

60. Leoni J, Gelmini S, Panzani G, Tanelli M (2024) A derivative, integral, and proportional features extractor for fault detection in dynamic processes. Eng Appl Artif Intell 128:107510. https://doi.org/10.1016/J.ENGAPPAI.2023.107510
61. Perez Bernal C, Salido MA, March Moya C (2025) Optimizing energy efficiency in unrelated parallel machine scheduling problem through reinforcement learning. Inf Sci 693:121674. https://doi.org/10.1016/J.INS.2024.121674
62. Li D, Zhao Y, Xi H (2024) A novel two-stage reinforcement learning framework for sustainable building energy management systems. J Build Eng 98:111475. https://doi.org/10.1016/J.JOBE.2024.111475
63. Al Sayed K, Boodi A, Sadeghian Broujeny R, Beddiar K (2024) Reinforcement learning for HVAC control in intelligent buildings: a technical and conceptual review. J Build Eng 95:110085. https://doi.org/10.1016/J.JOBE.2024.110085
64. Ajifowowe I, Chang H, Lee CS, Chang S (2024) Prospects and challenges of reinforcement learning-based HVAC control. J Build Eng 98:111080. https://doi.org/10.1016/J.JOBE.2024.111080
65. Liang R, Wang PH (2024) Enhancing energy efficiency in buildings, optimization method and building management systems application for lower CO_2 emissions. Energy 313:134054. https://doi.org/10.1016/J.ENERGY.2024.134054
66. Hernández JL, de Miguel I, Vélez F, Vasallo A (2024) Challenges and opportunities in European smart buildings energy management: a critical review. Renew Sustain Energy Rev 199:114472. https://doi.org/10.1016/J.RSER.2024.114472
67. Nagy Z, Henze G, Dey S, Arroyo J, Helsen L, Zhang X, Chen B, Amasyali K, Kurte K, Zamzam A, Zandi H, Drgoňa J, Quintana M, McCullogh S, Park JY, Li H, Hong T, Brandi S, Pinto G, Capozzoli A, Vrabie D, Bergés M, Nweye K, Marzullo T, Bernstein A (2023) Ten questions concerning reinforcement learning for building energy management. Build Environ 241:110435. https://doi.org/10.1016/J.BUILDENV.2023.110435
68. Ashouri M, Haghighat F, Fung BCM, Yoshino H (2019) Development of a ranking procedure for energy performance evaluation of buildings based on occupant behavior. Energy Build 183:659–671. https://doi.org/10.1016/J.ENBUILD.2018.11.050
69. Heti K, Silvanskaya G, Khambir V, Harkava V, Ivanchenko K (2025) The analysis of engineering adaptation to global challenges and strategic planning for the future. Sustain Eng Innov 7:137–146. https://doi.org/10.37868/SEI.V7I1.ID453
70. Wang X, Wang P, Huang R, Zhu X, Arroyo J, Li N (2025) Safe deep reinforcement learning for building energy management. Appl Energy 377:124328. https://doi.org/10.1016/J.APENERGY.2024.124328

Chapter 5
Evolving Zero Energy Building Practices and Their Role in Global Sustainability

Abbreviations

HVAC	Heating, ventilation, and air conditioning
NZEB	Net zero energy buildings
nZEB	Nearly zero energy buildings
PV	Photovoltaic
SDG	Sustainable development goal
WWR	Window-to-wall ratio
ZEB	Zero energy buildings

5.1 Introduction

The construction sector is ranked as number one in energy consumption and plays a major role in addressing climate change and resource depletion around the world. Therefore, the shift of the construction sector toward more sustainable practices has become essential. In this context, Zero Energy Buildings (ZEBs), also known as Net Zero Energy Buildings (NZEBs), have emerged as a strategy to tackle these challenges and guide the construction sector toward sustainability. These buildings are designed to produce energy from renewables equivalent to their annual demand. The evolution of ZEB practices is driven by advances in building technologies, materials and it is transforming architectural and engineering standards [1].

An overview of global energy consumption across major sectors including buildings, transport, and industry is shown in Table 5.1. It highlights the share of each sector in global energy use its consumption drivers and key sustainability initiatives underway.

Referring to Table 5.1, it was observed that buildings hold the number one position in terms of final energy use, followed by industry, and transportation in the third

B. Duraković, *Advancing Zero Energy Buildings*, Green Energy and Technology,
https://doi.org/10.1007/978-3-032-16601-2_5

Table 5.1 Global energy consumption by sector: key drivers, and sustainability focus

Sector	Energy use	Key drivers	Sustainability focus and sources	
Buildings	~ 34 – 37%	Urbanization, ACs, appliances	• Smart buildings, heat pumps • Energy-efficient appliances	[2, 3]
Transport	~ 20–29%	Motorization, logistics	• EV adoption and e-buses • Electrified mass transit	[3, 4]
Industry	~ 30–33%	Manufacturing, heavy industries	• Decarbonization with renewables • Efficiency programs	[3, 5]
Agriculture and other	~ 5–10%	Food demand, irrigation, agritech	• Solar irrigation and cold chains • Sustainable farming practices	[3]

place. Therefore, these sectors are in focus of global efforts to reduce emissions and promote sustainability. Figure 5.1 shows the distribution of energy consumption among dominant sectors in 2024. From the figure it can be seen that the largest consumers are buildings.

Based on the International Energy Agency's (IEA) Global Energy Review 2025 [6], global energy demand increased by 2.2% in 2024, driven by factors such as extreme temperatures, increased electrification, and digitalization. The buildings sector experienced a significant rise in electricity demand, nearly 60% of this overall growth due to increased demand for AC and cooling of data centers. The transportation demonstrated a modest increase due to the adoption of electric vehicles (EVs). Industry energy consumption rise is significant as well due to energy-intensive manufacturing processes [6].

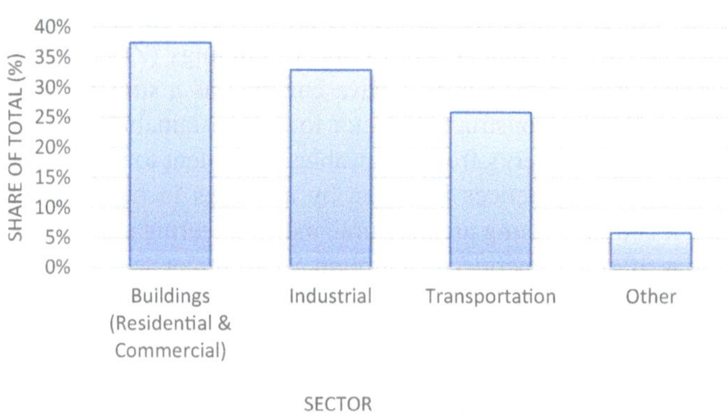

Fig. 5.1 Global energy consumption estimates by sector for 2024. *Author* based on data in [3]

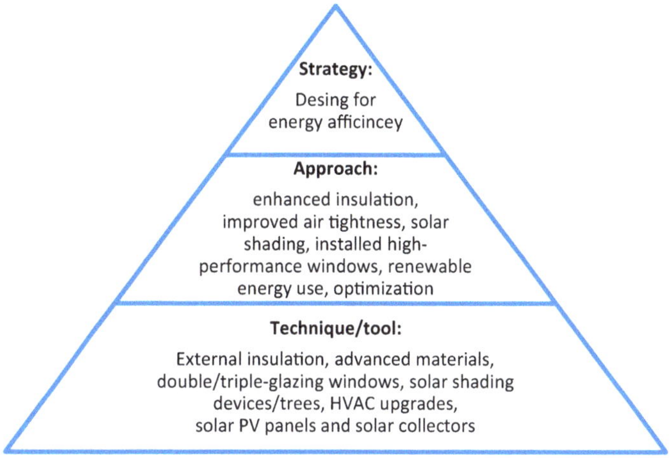

Fig. 5.2 Energy-efficient approaches and techniques

ZEBs were initially viewed as theoretical/experimental projects constrained by high costs and technological limitations. Advancement in building technologies and renewable energy systems, moved the ZEBs from experimental concepts to practical solutions. These solutions include a holistic design approach incorporating advanced insulation, high-performance windows, envelope airtight, and smart energy technologies. The integration of phase-change materials (PCMs), withing building structure (e.g. envelope) improves thermal inertia of the structure and reduces heating/cooling loads [7, 8]. Figure 5.2 depicts key energy-efficient approaches and techniques that form the foundation of sustainable building practices.

The primary ZEB renewable energy source is solar photovoltaic (PV) panels, but the advancements in battery storage, geothermal systems, and wind energy have expanded the possibilities for self-sustaining buildings. Recent digital tools, like energy simulation software, enable designers to optimize energy performance from the earliest stages of project development [9]. Despite economic and infrastructural challenges in developing countries, the ZEB potential is recognized to enhance energy security and support sustainable urban growth [10]. Tailored strategies that consider local climate, materials, and socioeconomic conditions are essential in adopting ZEBs globally.

The purpose of this chapter is to provide an overview of the key ZEB principles, practices and their importance in modern architectural design as well as to provide future direction for integrating ZEB concepts into new building projects.

5.2 Key ZEB Principles and Practices

Although the ZEB concept has been around for decades, there is no internationally aligned definition and pathway to implementation yet. Countries have entered in and out of discussions with different definitions and applications, usually at the residential buildings level. While a small number of ZEB case studies show great progress in achieving significant energy reductions, for now, most motivated governments have only taken a highly limited step, with a limited number of demonstration projects across countries where low-energy policies are more cohesive. The absence of a common global strategy has highlighted the need for more international cooperation and knowledge sharing [11].

At the heart of this architecture is the specific concept of energy performance and recovery that must be carried out, energy consumption and renewable energy production. Yet, setting terms such as "energy balance," "balance period," and which types of energy use to include in calculations is one of the problems. This includes the harmonization of renewable energy sources, demand-side management, thermal comfort, and the interaction of buildings with appropriate grid infrastructures. A target framework is necessary to ensure that ZEB targets are meaningful and comparable across projects and regions [12]. The goal of achieving ZEB standards is related to the more general goal of improving the energy performance of the entire building sector. Addressing barriers such as information gaps, limited resources, and insufficient policies is critical in energy demand reduction and improving building energy performance. These initiatives further serve to advance wider sustainability objectives, including UN Sustainable Development Goals (SDGs), particularly those focused on SDGs 11 and 13 [13].

5.2.1 Strategic Approaches to Achieving ZEB Standards

ZEB standards can be achieved through several specific strategies and technologies. The fundamental strategies and methodologies include dynamics like building envelope efficiency and optimization, passive vs. active heating, ventilation and air conditioning (HVAC) systems, renewable energy deployment and analysis, and geographic climates. Table 5.2 shows key strategic are with key insights, real-world examples, recommendations, and challenges.

The objective of *energy performance and holistic ZEB strategies* objectives is to reduce environmental impact through the active and passive strategies and collaboration between different stakeholders. Active and passive strategies are the most effective in reducing energy consumption. However, further research and a multidisciplinary approach is needed to establish ZEBs as a global standard [19].

The optimization of building envelopes is the first step to minimizing energy demand. A case study from s Swedish NZEB office building shows how the energy efficiency is improved by mitigation thermal bridges, the use of low U-value

Table 5.2 Strategic approaches in achieving ZEB standards

Focus area/ Strategy	Key insights	Examples	Recommendations	Challenges
Energy performance and holistic ZEB strategies	Energy performance is essential to minimizing environmental impact. Achieving ZEB involves active and passive solutions and collaboration among stakeholders [14]	General global perspective [15, 16]	• Use integrated design with multiple stakeholders • Combine passive and active energy strategies	Requires more research and international harmonization
Building envelopes and passive design optimization	Building envelope efficiency is the first step. Airtightness, thermal bridge mitigation, and high-performance windows are crucial to lowering energy demand [14]	Swedish net ZEB office building	• Increase airtightness- reduce thermal bridging • Use windows with low u-values	Even successful net ZEB buildings may not meet other strict standards (e.g., passivhaus)
Solar energy systems integration	Solar thermal and PV systems can cover 100% of heating and hot water needs, particularly in high irradiance regions. Solar integration is critical in both new builds and retrofits [17]	Portugal multi-family retrofit case	• Install solar PV and thermal systems- collaborate with local agencies for implementation-prioritize solar in sunny climates	Broader awareness and policy support needed for eu-wide adoption
Multi-strategy ZEB approaches	ZEB goals can be achieved via diverse methods combining envelope design, HVAC strategies (passive vs. Active), renewable energy integration, and context-specific solutions. Renovation assessment is also key to reaching nZEB and ZEB standards in existing buildings	Applicable to both new and retrofitted buildings globally	• Tailor solutions to climate and site- analyze renewable energy potential- consider passive/active HVAC trade-offs- assess deep retrofit potential	Context-specific design and retrofitting can be complex and require local adaptation [18]

windows, and improved air tightness [20]. A case study from Portugal demonstrated the importance of solar energy systems integration in a retrofit multi-family building. They used solar collectors and photovoltaic panels to cover complete energy demand for heating and hot water. The results showed that deep energy upgrades to achieve nZEB and ZEB standards resulted in energy reduction and carbon emissions within the 30-year analysis. The study further emphasizes the importance of working with local offices to increase awareness as well as conducting more research to facilitate a wider adoption ZEB standards in the EU [21–23].

5.2.2 ZEB Life Cycle and Cost Optimization

To achieve truly sustainable building, structures should be evaluated based on their entire life cycle, including the full environmental impact of the materials used in construction, throughout their operational life, and during post-use disposal. Sustainability with life cycle assessments is central role in the design process accounting operational and embodied energy. It addresses methods for determining cost-optimal energy performance levels as well as policy drivers to promote ZEBs.

The traditional ZEB definition focuses on operational energy (energy used while the building is in operation), Life Cycle ZEB considers both operational energy and embodied energy (the energy consumed during material production and recycling process). A life cycle approach has been suggested for a metric to inform building design, the Net energy ratio. This integrated approach calls for looking beyond building energy usage towards a full building life cycle to reach actual net-zero energy performance. This approach is picking up steam in building regulations, as the EU and UK are looking to achieve net-zero buildings [24].

In order to achieve ZEB standard, the interaction of energy performance and cost must be taken into consideration. Methodologies for determining cost-optimal energy performance levels have been a major focus of studies. A seven-step methodology for calculating these levels for residential buildings is defined by the Federation of European Heating, Ventilation and Air Conditioning Associations. If it is decided that primary energy consumption is cost-optimal, with a few simulations it can be shown, according to research studies. A recent comparison with an Estonian reference dwelling has proven that the optimum price levels are considerably lower than the current requirements. The EPBD recast directive sets out high-level requirements but needs to be transposed into national law which needs more specific statutory infrastructure. The introduction of accurate costing is critical towards reaching ZEB objectives, while the developed processes can be directed at the national stage in guiding ZEB energy performance decision [25]. Figure 5.3 show conceptual determination of cost-optimal energy performance for nearly zero energy buildings design. The procedure can be represented with a cost-optimal graph where the building energy consumption is on the horizontal axis, and the total cost is represented on the vertical axis.

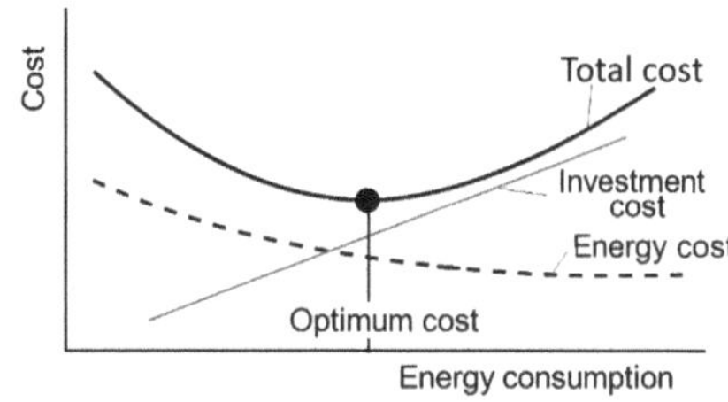

Fig. 5.3 Cost-optimal analysis for nearly zero energy buildings design

Cost incentives can be employed to better promote the use of ZEBs across different applications. One example of this type of approach is a reward-penalty system that rewards better use of renewable energy and penalizes lower levels of ZEB performance. Long-term benefits to building owners from ZEBs are substantial despite higher upfront costs to achieve levels above ZEB. So, it makes ZEB adoption financially attractive by means of a quadratic formula to adjust costs. Such a system provides a simple and effective approach to support the wider adoption of ZEBs [26].

5.3 Technological Innovations Driving ZEB Development

The pathway to zero energy buildings is highly dependent on the impact of integrated technologies/systems. Such technologies include enhanced building envelope, renewable energy generation (photovoltaics, solar thermal, wind, offshore wind, geothermal, etc.), more efficient HVAC systems, energy storage, smart energy management systems, and co-generation technologies. Enhancing the building envelopes is fundamental for reducing energy demand in ZEBs. Enhanced facades, windows, insulation, shading, daylighting systems, and sustainable building materials are good strategies for reducing energy demand.

Facade design and window design optimization are the most important in terms of energy efficiency. Some of the most important factors include window features, window-to-wall ratio (WWR), insulation thickness, and shading techniques. The ideal size of the window varies. Larger WWRs are preferrable for quad and quintuple glazing while smaller WWRs are preferred for double and triple glazing. The goal is to find an optimum balance between the insulation thickness and WWR.

Depending on the window size, outdoor shading systems are improving energy efficiency of the envelope, but these solutions significantly increase costs [27]. Particularly, shading devices have a significant effect on the cooling and lighting loads. Depending on climate zone and building orientation, it is necessary to establish proper guidelines for selecting appropriate window sizes and types of shading devices in early stage of the design process [28, 29].

Building materials and construction methods also affect energy efficiency. The growing use of bio-based materials, as well as the ability to use prefabrication techniques, is significant. As indicated by the number of papers published between 2015 and 2025 on bio-based materials, these materials provide alternative solutions, especially for single-family housing. It is expected that bio-based prefabrication will lead to quicker, cheaper, and greener construction [30–32].

The integration of renewable energy into existing buildings goes hand in hand with wider decarbonization goals. For example, installing solar PV on rooftops of buildings with the best angles can be obtained using optimization models. These models can assist building owners and stakeholders in effectively adopting renewable energy technologies. Policies related to land use policy promoting the integration of building PV systems and enhanced heat pump efficiency have potential for increasing energy savings [33–35].

New approaches in terms of HVAC systems include support with passive design (insulation and phase change materials), and heat pumps. Hybrid HVAC systems are particularly recommended for severe climates. It is observed that water-based air systems perform better than their air-based one. It can be expected that future systems will be combined with thermal energy storage to utilize renewable energy sources more effectively. Other studies suggest that nested hierarchical control approaches on increasing usability of thermal mass as a temporary thermal battery [36].

Passive design strategies are associated with net-zero energy buildings when combined with hybrid renewable energy systems. Various studies have focused on the use of Trombe walls for heating, underground cooling ducts for cooling, and solar hot water systems. They can dramatically reduce heating and cooling loads through passive strategies. When combined with hybrid renewable systems, including solar, wind turbines, batteries, and inverters, these passive elements can promote sustainability and economic feasibility even more [37, 38].

The transition to zero energy buildings is enabled through energy storage. This requires defining optimal storage capacities for heat, cold, and batteries. These integrated energy storage solutions with the building systems lead to decreased stress on the grid and improved functionality overall [39].

Emerging technologies such as machine learning and 3D modelling are helping to streamline design optimization. Simulation is a powerful tool for estimating energy use and emissions over the lifespan of a building. Novel technologies such as artificial intelligence (AI), deep learning, and the Internet of Things (IoT) have a lot of potential in making energy control more efficient in smart buildings [40]. The artificial intelligence-based frameworks can predict consumption, analyze data, detect inefficiencies, and even optimize usage via machine learning in real-time. In buildings with large heating and cooling loads like that of a hospital, co-generation systems (production of heat and electricity in a single process from a single fuel) hold potential to support ZEB targets. For heat storage, the combination of outer layers with PCMs can improve the performance of these systems [41, 42]. Energy modelling using artificial intelligence helps in finding designs with minimum energy consumption and even reduces carbon emissions from buildings [43].

The neighborhood or community scale expands the ZEB concept and can be a challenge and an opportunity. The ability to place zero-energy strategies at this level, looking at how urban form and location impact energy and transport. This includes energy sharing and the interaction between energy use in buildings and transportation. The durability of materials and the potential for future system performance degradation is critical in long run of ZEBs [44, 45].

5.4 Future of Building Design and Energy Efficiency

ZEBs are promising solutions to mitigate building sector pollution [46] and contributing to contributing to UN SDGs, particularly (SDG 7: Affordable and Clean Energy) and (SDG 13: Climate Action) [47]. These buildings are associated with clean energy and sustainable cities, which utilize renewable energy systems and recycling materials. In addition, sustainable finance mechanisms are also vital to support ZEBs project development and application, to ensure that financial research is aligned with the SDGs. Whereas financial research becomes more focused on green economy topics, the research must include some sustainable and climatic consideration at a broader stage [13, 48, 49].

In recent years, smart buildings have become popular in ZEB reaching, aiming to promote energy flexibility, renewable energy sources, and user interaction. Most of the today's building stock is likely to be occupied for decades, therefore smart retrofitting of the existing building stock will be imperative to adapt to a changing climate and grid. AI will play a significant role in this process but needs to be balanced against data privacy issues, energy consumption of AI systems themselves and fairness [50–52].

Reviews of progress in ZEB recommended development of regulations, policies, and incentives to strengthen ZEB efforts. These recommendations are intended to ensure that carbon–neutral building policies and future energy and emissions targets are successful. As the shift towards ZEB continues, government policy will likely include retrofitting existing buildings, integrating smart technologies and financing [53, 54].

From the challenges listed in Table 5.3 to the opportunities mentioned, all aspects will oiutline the development and implementation process of ZEBs. Highlighting the challenges that must be overcome and innovative opportunities available to further progress in ZEB design and implementation, it offers a snapshot of key themes that speak to general concepts of standardization, economic viability, technological integration, and climate responsiveness.

Referring to Table 5.3, a major challenge can be considered, lack of standardized definitions and performance metrics across regions. Another critical constraint is the high cost of advanced technologies and materials which require innovative incentive mechanisms and solutions. The integration of AI and IoT in the existing construction practices makes the process more complex and challenging.

ZEBs have become viable at the building scale, their implementation at the neighborhood or community scale needs to accommodate the specific nature of urban form, location, and energy mutualization as per the boundary condition of such systems [58–60]. All together these challenges provide an insight into evolving

Table 5.3 Key challenges and opportunities in advancing ZEBs

Aspect	Challenges	Opportunities	References
Standardization	Absence of global harmony in ZEB definitions and metrics	The global effort to standardize methodology and processes	[12]
Economic viability	High Implementation Costs of ZEB Technologies and Materials	Novel incentives (e.g., reward-penalty systems) for adoption	[26]
Technological Integration	Challenges in integrating new-age systems (AI, IoT) with traditional construction systems	Smart building solutions and energy management in emerging areas	[55–57]
Climate responsiveness	Hard to adapt the ZEBs to varying climates and advance conditions	Design conditioned for regionality through parametric and simulation-based technologies	[27]

balance between technological advancement, economic feasibility, and contextual adaptability in the global pursuit of sustainable building design.

5.5 Conclusion

This chapter investigated the role of integrated renewable energy systems, optimized building envelopes, and innovative financing mechanisms for energy neutrality. However, high initial costs and absent concrete definitions are hurdles to the global-level uptake of ZEBs. In addressing current challenges, the literature showed the importance of collective actions in policy making, technology development, and international standardization. Even as the ZEB sector evolves, long-term durability, scalability, and adaptability of ZEB across a wide range of climates, are potential areas of future research.

References

1. Zero-Energy Buildings. https://www.meegle.com/en_us/topics/sustainable-energy/zero-energy-buildings?utm_source=chatgpt.com. Last accessed 25 June 2025
2. Zhang S, Ma M, Zhou N, Yan J (2024) GLOBUS: global building renovation potential by 2070
3. Total Energy Annual Data—U.S. Energy Information Administration (EIA). https://www.eia.gov/totalenergy/data/annual/. Last Accessed 08 July 2025
4. Energy Agency I (2024) Global EV outlook 2024 moving towards increased affordability
5. Industry—Energy System—IEA. https://www.iea.org/energy-system/industry. Last Accessed 08 July 2025

6. Global Energy Review 2025
7. Reddy VJ, Ghazali MF, Kumarasamy S (2024) Advancements in phase change materials for energy-efficient building construction: a comprehensive review. J Energy Storage 81:110494. https://doi.org/10.1016/J.EST.2024.110494
8. Duraković B (2020) PCMs in building structure. In: PCM-based building envelope systems: innovative energy solutions for passive design. Springer, pp 63–87. https://doi.org/10.1007/978-3-030-38335-0_4
9. Chen L, Hu Y, Wang R, Li X, Chen Z, Hua J, Osman AI, Farghali M, Huang L, Li J, Dong L, Rooney DW, Yap PS (2023) Green building practices to integrate renewable energy in the construction sector: a review. Environ Chem Lett 22(2):751–784. https://doi.org/10.1007/S10311-023-01675-2
10. Jonasson J, Faith-Ell C, Carlman I, Englund O (2024) The environmental performance of zero-emission buildings in a fossil-free energy system. Energy Effic 17:1–17. https://doi.org/10.1007/S12053-024-10253-6/FIGURES/5
11. Panagiotidou M, Fuller RJ (2013) Progress in ZEBs—a review of definitions, policies and construction activity. Energy Policy 62:196–206. https://doi.org/10.1016/J.ENPOL.2013.06.099
12. Marszal AJ, Heiselberg P, Bourrelle JS, Musall E, Voss K, Sartori I, Napolitano A (2011) Zero energy building—a review of definitions and calculation methodologies. Energy Build 43:971–979. https://doi.org/10.1016/J.ENBUILD.2010.12.022
13. Hafez FS, Sa'di B, Safa-Gamal M, Taufiq-Yap YH, Alrifaey M, Seyedmahmoudian M, Stojcevski A, Horan B, Mekhilef S (2023) Energy efficiency in sustainable buildings: a systematic review with taxonomy, challenges, motivations, methodological aspects, recommendations, and pathways for future research. Energy Strat Rev 45:101013 (2023). https://doi.org/10.1016/J.ESR.2022.101013
14. Kang JE, Ahn KU, Park CS, Schuetze T, Rosen MA, Tieben H, Chelleri L, Ostermeyer Y, Wolfram M (2015) Assessment of passive vs. active strategies for a school building design. Sustainability 7:15136–15151. https://doi.org/10.3390/SU71115136
15. Buzatu R, Ungureanu V, Ciutina A, Gireadă M, Vitan D, Petran I (2021) Experimental evaluation of energy-efficiency in a holistically designed building. Energies 14:5061. https://doi.org/10.3390/EN14165061
16. Iyer-Raniga, U (2019) Zero energy in the built environment: a holistic understanding. Appl Sci 9:3375. https://doi.org/10.3390/APP9163375
17. Neto RC, Ferreira X, Silva CS (2023) Phase change materials in a hybrid solar thermal/photovoltaic energy storage system for a residential house. Sustain Build 6:3. https://doi.org/10.1051/SBUILD/2023003
18. Citadini de Oliveira C, Catão Martins Vaz I, Ghisi E (2024) Retrofit strategies to improve energy efficiency in buildings: an integrative review. Energy Build 321:114624. https://doi.org/10.1016/J.ENBUILD.2024.114624
19. Belussi L, Barozzi B, Bellazzi A, Danza L, Devitofrancesco A, Fanciulli C, Ghellere M, Guazzi G, Meroni I, Salamone F, Scamoni F, Scrosati C (2019) A review of performance of zero energy buildings and energy efficiency solutions. J Build Eng 25:100772. https://doi.org/10.1016/J.JOBE.2019.100772
20. Berggren B, Wall M, Flodberg K, Sandberg E (2012) Net ZEB office in Sweden—a case study, testing the Swedish Net ZEB definition. Int J Sustain Built Environ 1:217–226. https://doi.org/10.1016/J.IJSBE.2013.05.002
21. Silva SM, Mateus R, Marques L, Ramos M, Almeida M (2016) Contribution of the solar systems to the nZEB and ZEB design concept in Portugal—energy, economics and environmental life cycle analysis. Sol Energy Mater Sol Cells 156:59–74. https://doi.org/10.1016/J.SOLMAT.2016.04.053
22. Ahsan Kabir M, Hasan MM, Hossain T, Ahnaf A, Monir H (2024) Sustainable energy transition in Bangladeshi academic buildings: a techno-economic analysis of photovoltaic-based net zero energy systems. Energy Build. 312:114205. https://doi.org/10.1016/J.ENBUILD.2024.114205

23. Makhloufi AW, Louafi S (2024) Optimising building performance for a resilient future: a multi-objective approach to net zero energy strategies. Energy Build. 324:114869. https://doi.org/10.1016/J.ENBUILD.2024.114869
24. Hernandez P, Kenny P (2010) From net energy to zero energy buildings: defining life cycle zero energy buildings (LC-ZEB). Energy Build 42:815–821. https://doi.org/10.1016/J.ENBUILD.2009.12.001
25. Kurnitski J, Saari A, Kalamees T, Vuolle M, Niemelä J, Tark T (2011) Cost optimal and nearly zero (nZEB) energy performance calculations for residential buildings with REHVA definition for nZEB national implementation. Energy Build. 43:3279–3288. https://doi.org/10.1016/J.ENBUILD.2011.08.033
26. Lu Y, Zhang XP, Li J, Huang Z, Wang C, Luo L (2019) Design of a reward-penalty cost for the promotion of net-zero energy buildings. Energy 180:36–49. https://doi.org/10.1016/j.energy.2019.05.049
27. Thalfeldt M, Pikas E, Kurnitski J, Voll H (2013) Facade design principles for nearly zero energy buildings in a cold climate. Energy Build 67:309–321. https://doi.org/10.1016/J.ENBUILD.2013.08.027
28. Tzempelikos A, Athienitis AK (2007) The impact of shading design and control on building cooling and lighting demand. Sol Energy 81:369–382. https://doi.org/10.1016/J.SOLENER.2006.06.015
29. Gupta M, Zala P, Gupta S, Varshney S (2024) Solar concentration based indoor daylighting system to achieve net zero sustainable buildings. Energy Build 321:114662. https://doi.org/10.1016/J.ENBUILD.2024.114662
30. Sutkowska M, Stefańska A, Vaverkova MD, Dixit S, Thakur A (2024) Recent advances in prefabrication techniques for biobased materials towards a low-carbon future: from modules to sustainability. J Build Eng 91:109558. https://doi.org/10.1016/J.JOBE.2024.109558
31. Firoozi AA, Firoozi AA, Oyejobi DO, Avudaiappan S, Flores ES (2024) Emerging trends in sustainable building materials: TECHNOLOGICAL innovations, enhanced performance, and future directions. Results Eng 24:103521. https://doi.org/10.1016/J.RINENG.2024.103521
32. Figueiredo A, Almeida RMSF, Vicente R, Ferreira VM (2024) Impact of heating strategies towards energy-efficient buildings. Energy Build. 315:114287. https://doi.org/10.1016/J.ENBUILD.2024.114287
33. Shimoda Y, Sugiyama M, Nishimoto R, Momonoki T (2021) Evaluating decarbonization scenarios and energy management requirement for the residential sector in Japan through bottom-up simulations of energy end-use demand in 2050. Appl Energy 303:117510. https://doi.org/10.1016/J.APENERGY.2021.117510
34. Hassan AA, El-Rayes K (2021) Optimizing the integration of renewable energy in existing buildings. Energy Build 238:110851. https://doi.org/10.1016/J.ENBUILD.2021.110851
35. ALRubeei IRN, Idi SN, Hussein Alsammak IL, AlRikabi HT, Mutar HA, Alaidi AHM (2025) Using artificial intelligence for enhancement of solar cell efficiency in the of Iraq. Herit Sustain Dev 7:197–210. https://doi.org/10.37868/HSD.V7I1.1067
36. Aljashaami BA, Ali BM, Salih SA, Alwan NT, Majeed MH, Ali OM, Alomar OR, Velkin VI, Shcheklein SE (2024) Recent improvements to heating, ventilation, and cooling technologies for buildings based on renewable energy to achieve zero-energy buildings: a systematic review. Results Eng 23:102769. https://doi.org/10.1016/J.RINENG.2024.102769
37. Sady H, Rashidi S, Rafee R (2024) Towards a net-zero-energy building with smart control of Trombe walls, underground air ducts, and optimal microgrid composed of renewable energy systems. Energy 294:130703. https://doi.org/10.1016/J.ENERGY.2024.130703
38. Santamouris M, Kolokotsa D (2013) Passive cooling dissipation techniques for buildings and other structures: the state of the art. Energy Build. 57:74–94. https://doi.org/10.1016/J.ENBUILD.2012.11.002
39. Medved S, Domjan S, Arkar C (2021) Contribution of energy storage to the transition from net zero to zero energy buildings. Energy Build 236:110751. https://doi.org/10.1016/J.ENBUILD.2021.110751

40. Jain R, Bakare YB, Alaric JS, Balam SK, Pattanaik B, Ayele TB, Nalagandla R (2023) Optimization of energy consumption in smart homes using firefly algorithm and deep neural networks. Sustain Eng Innov 5:161–176. https://doi.org/10.37868/SEI.V5I2.ID210
41. Alijoyo FA (2024) AI-powered deep learning for sustainable industry 4.0 and internet of things: enhancing energy management in smart buildings. Alex Eng J 104:409–422. https://doi.org/10.1016/J.AEJ.2024.07.110
42. Assareh E, Keykhah A, Cao Nhien L, Ghodrat M, Firoozzadeh M, Lee M (2024) A proposal on a co-generation system accompanied with phase change material to supply energy demand of a hospital to make it a zero energy building (ZEB). Energy Build. 318:114478. https://doi.org/10.1016/J.ENBUILD.2024.114478
43. Gan VJL, Lo IMC, Ma J, Tse KT, Cheng JCP, Chan CM (2020) Simulation optimisation towards energy efficient green buildings: current status and future trends. J Clean Prod 254:120012. https://doi.org/10.1016/J.JCLEPRO.2020.120012
44. Marique AF, Reiter S (2014) A simplified framework to assess the feasibility of zero-energy at the neighbourhood/community scale. Energy Build. 82:114–122. https://doi.org/10.1016/J.ENBUILD.2014.07.006
45. Danza L, Belussi L, Guazzi G, Meroni I, Salamone F (2018) Durability of technologies in the keeping of ZEB's performances. Energy Procedia. 148:138–145. https://doi.org/10.1016/J.EGYPRO.2018.08.041
46. Ma Z, Arıcı M, Sun Y, Singh S, Shahsavar A (2024) Towards net-zero energy/emission buildings for sustainable development. Energy Sustain Dev 80:101448. https://doi.org/10.1016/J.ESD.2024.101448
47. Olabi AG, Elsaid K, Obaideen K, Abdelkareem MA, Rezk H, Wilberforce T, Maghrabie HM, Sayed ET (2023) Renewable energy systems: comparisons, challenges and barriers, sustainability indicators, and the contribution to UN sustainable development goals. Int J Thermofluids 20:100498. https://doi.org/10.1016/J.IJFT.2023.100498
48. Olabi AG, Shehata N, Issa UH, Mohamed OA, Mahmoud M, Abdelkareem MA, Abdelzaher MA (2025) The role of green buildings in achieving the sustainable development goals. Int J Thermofluids 25:101002. https://doi.org/10.1016/J.IJFT.2024.101002
49. Su Y, Lucey BM, Jha AK (2024) Finance research and the UN sustainable development goals—an analysis and forward look. Res Int Bus Finance 71:102463. https://doi.org/10.1016/J.RIBAF.2024.102463
50. Al Dakheel J, Del Pero C, Aste N, Leonforte F (2020) Smart buildings features and key performance indicators: a review. Sustain Cities Soc 61:102328. https://doi.org/10.1016/J.SCS.2020.102328
51. Olawade DB, Wada OZ, David-Olawade AC, Fapohunda O, Ige AO, Ling J (2024) Artificial intelligence potential for net zero sustainability: current evidence and prospects. Next Sustainability. 4:100041. https://doi.org/10.1016/J.NXSUST.2024.100041
52. Al-Tamimi SA, Anssari MA, Al-Tameemi MD (2025) The impact of applying artificial intelligence-based systems on human resource costs in the light of environmental challenges. Herit Sustain Dev 7:245–256. https://doi.org/10.37868/HSD.V7I1.1091
53. Zhang S, Wang K, Xu W, Iyer-Raniga U, Athienitis A, Ge H, woo Cho D, Feng W, Okumiya M, Yoon G, Mazria E, Lyu Y (2021) Policy recommendations for the zero energy building promotion towards carbon neutral in Asia-Pacific Region. Energy Policy 159:112661. https://doi.org/10.1016/J.ENPOL.2021.112661
54. Economidou M, Todeschi V, Bertoldi P, D'Agostino D, Zangheri P, Castellazzi L (2020) Review of 50 years of EU energy efficiency policies for buildings. Energy Build. 225:110322. https://doi.org/10.1016/J.ENBUILD.2020.110322
55. Aste N, Del Pero C, Leonforte F, Ounis S, Abdelrahim AMG (2024) The role of photovoltaic technology in achieving net zero energy target in residential buildings. Energ Strat Rev 55:101543. https://doi.org/10.1016/J.ESR.2024.101543
56. Jing T, Zhao Y (2024) Optimizing energy consumption in smart buildings: a model for efficient energy management and renewable integration. Energy Build. 323:114754. https://doi.org/10.1016/J.ENBUILD.2024.114754

57. Heti K, Silvanskaya G, Khambir V, Harkava V, Ivanchenko K (2025) The analysis of engineering adaptation to global challenges and strategic planning for the future. Sustain Eng Innov 7:137–146. https://doi.org/10.37868/SEI.V7I1.ID453
58. Shaad M, Gupta P, Parvez M, Khan O, Kumari P (2024) Recent trends and future potential of sustainable energy efficient materials for commercial buildings. Handb Emerg Mater Sustain Energy 725–739. https://doi.org/10.1016/B978-0-323-96125-7.00011-3
59. Ohene E, Chan APC, Darko A (2022) Review of global research advances towards net-zero emissions buildings. Energy Build. 266:112142. https://doi.org/10.1016/J.ENBUILD.2022.112142
60. Shadram F, Mukkavaara J (2024) Investigating the potential for realizing life cycle net-zero energy buildings in Europe using multi-objective optimization. Energy Rep 12:5648–5670. https://doi.org/10.1016/J.EGYR.2024.11.042

Chapter 6
Lesson Learned from ZEB Projects Worldwide

6.1 Introduction

The paradigm shift in building design and construction methodology was significantly created due to the urgent need for reduced energy consumption and mitigating climate changes [1–3]. Within this frame, Zero Energy Buildings (ZEBs) have been recognized as an important driver towards sustainable development.

This chapter discusses several worldwide case studies with an idea of diversified applications of ZEB principles in residential, commercial, and institutional buildings: from single-family houses to multi-family dwellings, from large-scale urban developments to campus projects through a wide range of energy and design scales [3, 4]. The chapter further discusses major lessons in terms of best practices, innovative solutions, and success stories with implementation that have driven the ZEBs' adoption worldwide. Although highly promising, the field still faces several challenges that ZEBs must address including construction and operation, such as high upfront costs, complexities of technical implementation, and constantly evolving regulatory frameworks. Using real case studies, this chapter identifies some of these barriers that stakeholders can work to address and advance the adoption of ZEB [4].

6.2 Residential ZEBs

It remains one of the essential parts of the built environment worldwide, consuming a substantial share of energy and producing a significant percentage of total anthropogenic greenhouse gas emissions. Increased urbanization coupled with a rise in populations has raised a growing need for reassessment of the approach toward design and operation of residential structures [5, 6]. These ZEBs are the new key-in-hand solution to the problem, as they allow these houses to meet their energy demands themselves in a very sustainable way while optimizing their reduced footprint on the

B. Duraković, *Advancing Zero Energy Buildings*, Green Energy and Technology,
https://doi.org/10.1007/978-3-032-16601-2_6

environment [6, 7]. In residential ZEBs, there will be two important but different environments to which energy-efficient design and renewable-energy technologies will have to be adapted: single-family homes and multifamily units.

Single-family ZEBs have long served as test beds for new energy solutions. Those were often test beds for advanced materials, passive design strategies, and renewable energy systems [8]. The adoption of several measures contributing to efficiency in conjunction with on-site energy generation that balances energy supply and demand is one of the common features of single-family ZEBs [5–7]. Minimization of heat losses and gains for high-performing building envelopes is directly enabled through insulation, airtight construction, and windows that are energy efficient. Further contribution comes from enabling passive design methodologies in order to provide maximum natural light utilization and achieve thermal comfort, optimizing window placement and integration of shades, disabling energy over-consumption through use of mechanical systems.

On-site community renewable energy systems include primarily solar photovoltaic panels, which supply clean energy to residences, often complemented with a battery storage system to assure energy delivery at any instant of demand spikes or when the sun is not bright enough [7, 9].

Multi-family dwellings include both challenges and opportunities characteristic to the implementation of ZEB. Allowing several families or individual owners to be part of the same structure complicates the management of energy use because of shared infrastructures and the greater number of occupants [10]. The advantages are also noticed when these systems are integrated within the urban fabric, as users often have combined spaces and use of resources, furthered through centralized systems such as geothermal heating and cooling systems or collective solar photovoltaic arrays looking at the maximization of energy optimization across units. The newly established energy-sharing model epitomized by microgrids, uses operational efficiency to position buildings to balance energy supply and demand [11, 12].

Advanced smart building technologies enhance energy efficiency in multifamily ZEBs by continuously monitoring energy consumption using energy management systems and sensors. Worldwide set case studies showed potential of ZEBs to transform building construction fields into suburban and rural contexts [13]. Such buildings have been able to save energy while offering comfort to occupants at affordable costs. Quite often, they demonstrate how to adapt ZEB principles to local climates and cultural preferences using the available regional materials and renewable energy sources. Similarly, in the densely populated urban areas, the ZEB projects in multifamily settings have their ways through which ZEB solutions can scale up to accommodate more occupants via multi-family, multi-storey apartments with roof-integrated solar panels, energy efficient HVAC, and social areas reinforcing an energy conscious lifestyle [13, 14].

Despite their promise, residential zero-energy buildings still face numerous challenges, many of which still have to be overcome to enable their widespread diffusion [14, 15]. Rather, notable barriers still exist, largely in the form of high up-front costs prohibitive to poor families and to developers with limited capital. While long-term economic benefits from reduced energy consumption and minimal maintenance are long-run economic dividends that often exceed the front-end costs, such financial requirements can prevent many potential candidates from being able to participate. Spatial limitations certainly exist, particularly in the metropolitan region, insomuch as space may be limiting regarding installing sufficient solar panels or any other viable renewable energy production system [14, 15]. All these situations can be exacerbated by the so-called regulatory obstacles, notably the outdated building codes and zoning ordinances, which do seriously inhibit innovation and progress. Besides, limited awareness and capacities of builders, architects, and homeowners will make the dissemination of ZEB principles very slow and hence create a need for education and capacity-building programs [15, 16].

However, lessons learned from existing residential ZEBs serve as valuable insights to surmount these challenges. Early involvement of the ZEB goals in the design process allows the interaction of all building components-envelope and energy systems-toward achieving a net-zero energy performance. This collaboration among the stakeholder group-homeowners, tenants, and developers-creates something similar to a shared commitment to energy efficiency and sustainability [16]. To compensate for high investment costs and making access feasible to more people, in some countries there are available innovative financing models including government subsidies, green loans, or performance-based incentives [17, 18]. This evidence for residential ZEBs helps the countries to bring forward the future where sustainable housing is the rule rather than the exception [18, 19], and can establish the foundation to continuous innovation, collaboration, and supportive policies around the world [19].

6.3 Commercial and Institutional ZEBs

The shift towards zero energy buildings has increasingly expanded into the commercial and institutional domains. Larger in size and serving various purposes, these buildings possess different challenges and opportunities towards net-zero energy efficiency. Commercial and institutional ZEBs are very important due to their general contribution to lowering the overall energy footprint in an urban environment, because they often represent the most energy-consuming parts of the built environment. They host workplaces, education facilities, health facilities, retail outlets, and several other important fields of human use where people spend a considerable amount of their day [20, 21].

Except for residential ZEBs that are for living areas, commercial and institutional buildings have highly complicated demands due to their size and functionality in relation to lighting, heating, cooling, and operation energy use. Therefore, it requires new approaches in design and advanced energy management, possibly with integrated on-site renewable energy systems to achieve net-zero status [22]. The chapter splits them into different but interrelated categories of commercial and institutional ZEBs and demonstrates the use of energy-efficient principles and sustainable technologies across non-residential spaces [23, 24].

6.3.1 Commercial ZEBs

Commercial buildings are broadly categorized into offices, retail stores, hospitality, and mixed-use, among others. These buildings are very important in modern economies, and they are the most intensive users of energy due to continuous lighting, HVAC systems, and special equipment. A drive toward commercial ZEBs is thus a solution to high energy demand that may provide a pathway towards reducing operational costs while significantly mitigating environmental impacts [23, 24]. Both passive and active energy solutions must be combined into a single design and implementation process for the commercial ZEBs. It thus requires the optimization of passive design strategies, such as building orientation, natural daylighting, and high-performance building envelopes that minimize baseline energy use. Simultaneously, it involves energy-efficient HVAC systems, advanced lighting, and smart energy management technologies as the core of operational efficiency. Large-format renewable energy installations have been integrated into many commercial net-zero energy buildings: examples include rooftop solar arrays and wind turbines [24, 25].

An example of commercial ZEBs that has usually been described as one of the greenest commercial buildings on the planet is the Bullitt Center in Seattle [26]. In fact, it demonstrates how ZEB principles can be applied in an office space without sacrificing functionality or comfort. It also has an efficient façade, with operable windows that permit natural ventilation [27, 28]. Additionally, a rooftop-based photovoltaic system and rainwater harvesting system reduce dependence from outside resources. Overall, these features make the achievement of net-zero energy performance and provides a productive environment for the occupants [28, 29]

However, commercial ZEBs also face significant challenges. Advanced development and investment costs are usually associated with advanced ventilation technologies and renewable energy systems; therefore, developers and investors, especially in speculative real estate markets, have usually avoided sustainability over profitability [30]. The integration of large-scale commercial projects is very complex and entails a lot of coordination among architects, engineers, contractors, and tenants in integrating ZEB principles. With such challenges, the long-term economic benefits of reduced operating costs and increased value will usually offset the initial investment, making a commercially viable ZEB a safe incentive for advanced investors [30, 31]. Table 6.1 summarizes performance metrics for some commercial ZEBs in various climate zones, showing key energy strategies used to achieve net-zero goals in different environmental conditions.

From Table 6.1 can be observed that local climate conditions influence energy use patterns and design strategies. Climate specific energy metrics are tailored for each case demonstrating passive design, airtight envelopes, and renewable energy integration. These examples show the importance of climate-responsive design in achieving ZEB goals [36, 37].

Table 6.1 Examples of energy performance of commercial ZEBs in different climates

Climate example	Energy performance	Remarks	References
Cold and dry climate, Idaho, USA	Whole-building energy use intensity (EUI) was 34.2 kWh/(m^2 y)	Gives end-use breakdown; a rare example with measured data for commercial ZEBs	[32]
Hot/wet summers, moderately cold winters, Calolziocorte, Italy	Base-case: 286 kWh/(m^2 y) average commercial building in Europe; With improvements: down to ~ 78.2 kWh/(m^2 yr); in best case (case-3) ~ 28.93 kWh/(m^2 yr); with onsite energy production case-4, closer to ~ 90% reduction	Clear demonstration of reducing energy demand via passive envelopes, airtightness, renewable supply etc	[33]
Tropical, South Jakarta, Indonesia	Energy consumption split: AC uses ~ 53% of electricity; electric equipment ~ 36% (lighting etc.) of remaining	Shows which loads dominate in tropical commercial building; useful for identifying where interventions matter most	[34]
Hot desert climate, Egypt	After applying reduction steps, consumption ~ 28 kWh/(m^2 y) before renewables; then installed PV + micro wind to cover needs, resulting in net zero version	Shows methodology towards ZEB in a commercial/public office context; simulation plus renewables	[35]

6.3.2 *Institutional ZEBs*

Institutional buildings, such as schools, universities, hospitals, and government facilities hold important places in the structure of society. Such kinds of buildings are mostly dependent on public funding and need to set leading examples for sustainable and resource-efficient practices. Institutional ZEBs have gained much focus because of their ability to cut operational costs, decrease carbon footprints, and offer environmental governance toward meeting the specific needs of their users [38]. Energy needs in institutional ZEBs vary significantly depending on their purpose. For example, hospitals require continuous operation due to 24/7 climate control, high-power medical equipment and extensive lighting systems. Schools and universities have more variable energy demands influenced by the educational calendar. Therefore, designing institutional ZEBs required specific attention to their unique operational needs [39, 40]. Table 6.2 shows selected journal case studies of institutional ZEBs across different climates zones spanning from hot-summer/cold-winter, and Mediterranean climates. Measured or simulated energy performance metrics such as Energy Use Intensity (EUI) or primary energy consumption are presented in the table and include retrofit or new buildings.

Table 6.2 Examples of energy performance of institutional ZEBs in different climates

Climate	Energy performance	Remarks	Article
Mild climate with warm, dry summers, Valladolid, Spain	Primary energy consumption ~ 67 kWh/(m^2 y); renewable generation ~ 121 kWh/(m^2 y); Renewable Energy Ratio (RER) ~ 0.66	Good example of institutional building achieving more renewables than certain base reference demands; shows "balance" between consumption and generation	[41]
Hot summer/cold winter zones, China	Reports consumption and emissions for several NZEBs in those climate zones; specific EUI values and carbon data per building are given	Allows comparison of how institutional buildings perform under challenging climates; useful for benchmarking	[42]
Mediterranean, Istanbul, Turkey	Current energy use baseline: ~ 225 TOE (tons of oil equivalent) for the building; with recommended interventions for NZEB/nZEB transition (TOE reduction etc.)	Large-scale institutional building; focus on retrofit/ performance improvement rather than new build	[43]
Mediterranean, Izmir, Turkey	Best scenario reduced net primary energy consumption by ~ 96.08% to 7.45 kWh/(m^2 y) vs. reference case	Though residential, provides insight into what aggressive retrofit plus renewables can achieve; useful benchmark for institutional building aspirations	[44]

Referring to Table 6.2, several patterns can be observed. For instance, buildings in mild climates may achieve relatively moderate primary energy consumption of $\approx$ 67 kWh/(m^2 y) while generating even more renewable energy of $\approx$ 121 kWh/(m^2 y), giving a renewable energy ratio around 0.66. In contrast, climates with hot summers and cold winters are more challenging in terms of energy demands and maintaining low emissions, while the mediterranean climates demonstrated significant improvements in reducing net primary energy consumption.

Institutional ZEBs also face challenges somewhat different from their commercial counterpart. Usually, public institutions must operate within tight budgetary allocations, which consequently limit the availability of funds for any upfront investment in sustainable technologies. In addition, such buildings have to meet strict standards for safety, accessibility, and functionality. These make it quite tricky to incorporate principles of ZEBs in them. To overcome such challenges, strong political will, innovative financing mechanisms, and leadership commitment to sustainability by both institutional leaders and policy are required [45–47].

The advantages of institutional ZEBs are not confined to energy savings alone: such buildings also act like living laboratories for sustainability by imparting various educational lessons to students, staff, and the community. By demonstrating the feasibility of net-zero energy performance, institutional zero energy buildings build public confidence in renewable energy technologies and spur further diffusion of these technologies into all building sectors [45, 48]. Table 6.3 presents the comparison between commercial and institutional ZEBs across multiple aspects such as energy, technologies utilized and the major challenges.

Based on Table 6.3, it can be concluded that the energy utilization of both commercial and institutional ZEBs have an aim of providing highly efficient means of energy generation, storage and performance of advanced systems to avoid possible risks.

Table 6.3 Comparison between commercial and institutional ZEBs across multiple aspects

Aspect	Commercial ZEBs	Institutional ZEBs
Energy use	• Typically designed to save 50–70% of energy compared to conventional buildings [49] [50]	• Energy needs vary significantly depending on the type of institution (e.g., hospitals require more energy for medical equipment) [49]
	• Aim to produce enough renewable energy to meet annual consumption requirements [51, 52]	• Focus on balancing energy consumption with renewable production, often with a goal of net-zero performance [51, 52]
	• Use of advanced energy management systems to optimize energy consumption and generation [53]	• May incorporate energy recovery systems and efficient HVAC tailored for specific operational needs [53]
Challenges	• High upfront costs associated with advanced technologies and renewable systems can deter investment [50, 54]	• Budget constraints in public institutions limit initial investments in sustainable technologies [55]
	• Complexity in integrating multiple stakeholders (architects, engineers, contractors) for design and implementation [56, 57]	• Must comply with strict safety, accessibility, and functionality standards, complicating ZEB integration [58, 59]
	• Market demand and regulatory pressures can influence development timelines and feasibility [60, 61]	• Need for strong political will and innovative financing mechanisms to support sustainability initiatives [62, 63]
Technologies Used	• Incorporate a variety of renewable energy sources such as solar panels and wind turbines [53, 64, 65]	• Utilize specialized technologies like geothermal heating/cooling systems, especially in hospitals and schools [53, 64–66]
	• Advanced building management systems that include smart technologies for monitoring and optimizing energy use [67]	• Often feature educational components that demonstrate sustainability practices to the community and students [68]
	• Focus on passive design strategies such as natural ventilation and high-performance building envelopes [67]	• Emphasize unique operational needs of institutions, requiring tailored solutions for energy efficiency [69]

The challenges that are faced are related to high initial cost, thereby establishing a proper way or means of funding is a major setback for such systems [50, 54, 55]. Regarding the technology, it can be stated that shifting towards renewables such as solar, wind, hydro provides highly potent way of producing electricity, however, the future systems would require highly optimized energy storage so that power factor is as close to 1 as possible [53, 64–66]

The major challenge for single and multi-ZEB households is the initial investment to fully replace the conventional way of energy supply. This challenge must be tackled to circumvent the transitional period and in the case of minor households, and governments may provide certain tax reductions and benefits for the people or communities willing to switch towards renewables [46, 47, 70].

6.4 Large Scale ZEBs

Large-scale ZEBs have become one of the paradigmatic solutions in view of growing challenges due to urbanization. Large-scale ZEBs transcend single buildings to campuses, integrated townships, and to urban regular developments, which present tremendous opportunities combined with complex challenges. These projects target integrated systems where energy-efficient buildings interact among themselves and with renewable energy infrastructures to achieve net-zero energy performance at community or district levels. Being able to implement energy consumption and sustainability on a far larger scale than previous examples can make these urban ZEB developments and campus projects scalable models for reducing carbon footprints while nurturing energy resilience and improving the quality of life for their inhabitants [56, 71].

6.4.1 Urban ZEB Projects

Urban ZEBs represent a visionary, out-of-the-box review of conventional paradigms of city planning and architectural design. In other words, conceived to achieve a net-zero energy performance at the neighborhood or district levels, it integrates several building types such as commercial, institutional into a single unit with shared renewable energy systems, smart grids of the state of art, and efficient energy storage in view of maximizing sustainability and energy efficiency in the new urban area [72].

Such ZEBs will require sensitive thinking and planning approaches in terms of energy efficiency and the best use of resources if the urban developments in them are to succeed. In brief, shared district energy systems are where many buildings can be provided with heating, cooling, and electricity that might be provided by renewable systems. These will be combined heat and power facilities, geothermal systems, or disseminated solar installations installed in an urban ZEB development [73].

Passive design strategies for each development include green roofs, urban vegetation, and energy-efficient building facades. Optimization of densities and placing buildings with a view to maximum solar exposure comprises the advanced practice of urban zoning so that energy production and consumption within the district can be balanced [74].

This can be clearly seen in the example of Masdar City in Abu Dhabi, which was touted to be one of the world's most sustainable cities. The city incorporates renewable energy systems, such as solar power and wind turbines, into the efficient building design in an attempt to achieve a net near-zero energy footprint. The city has various updated planning methods of the time, including shaded roadways, and wind channeling to reduce cooling needs of the city's hot desert climate [75].

6.4.2 *Campus Projects*

Large-scale ZEB principles could be put into practice in university and corporate campuses. These practically are self-sustaining communities with huge demands on energy consumption, diversification of functionality including education, research, housing, and facilities for recreation. In that respect, campus ZEB projects offer suitable opportunities to develop innovative energy solutions and provide a model to the students, faculty members, and visitors.

One of the important features of ZEBs on campus is the ability to integrate different building types and functions into an integral energy system. This might be achieved through the use of microgrids that allow campuses to locally produce, store, and distribute energy. Microgrids increase the resilience of energy by permitting campuses to continue functioning from the larger grid during outages or in cases of other emergencies [76, 77]. The other defining feature of these campus ZEBs was the integration of renewable technologies, such as solar farms, wind turbines, or biomass generators. Most often, this was coupled with energy storage solutions, comprising battery banks and thermal storage tanks, in order to provide a consistent supply of energy [78]. Advanced energy management systems are crucial for optimally using energy in the buildings on campus by balancing demand and supply with real-time data [79]. Passive design strategies are also critical in campus ZEB projects. For example, the orientation of campus buildings is usually planned in a way to maximize solar gain during winter and minimize heat gain during summer, which reduces the need for mechanical heating and cooling. Further energy efficiency is provided by green roofs, natural ventilation systems, and daylighting techniques [80].

A state-of-the-art example was when West Village at the University of California, Davis, was constructed to be the largest planned zero-energy community in the United States. It combines energy-efficient building design with a 4-megawatt solar farm and advanced energy management systems, effectively achieving net-zero energy performance on campus. For example, such a project-on-campus student and faculty housing, retail, or recreational space-demonstrates an application of the principles of

ZEB in a multi-functional development area [81]. Another fine example would be the integration of the concepts of the Zero Energy Building into the campus buildings by the Nanyang Technological University, Singapore, under its sustainability mantle. For example, at the Nottingham Trent University, the School of Art, Design, and Media has a green roof, with advanced cooling systems reducing energy consumption by about 30%. The campus is powered by a variant of solar panels mashed together with energy-efficient technologies in order to meet its energy needs [82].

Benefits of campus ZEBs are reflected in reduced energy consumption, improved energy resiliency, and enhanced environmental performance. They often serve as living learning labs on campus for students and staff, providing hands-on experience with sustainable design and energy management. Most projects are enabled through some source of funding and most or all enhance recognition for the institution [83]. On the other hand, there are still some other major roadblocks on the path to reaching Campus ZEBs: large upfront investments in renewable energy systems and retrofits of existing buildings. Increased system design and implementation complexity reflects the diversity of campus energy uses, ranging from laboratories to sport facilities. Many of these challenges can be overcome through active institutional leadership, innovative financing, and collaboration with partners outside the institution [83, 84].

6.5 Lessons Learned from Successful ZEB Implementation

The road to the realization of zero energy buildings is filled with turning points, challenges, and lessons learned. Such insights are relevant for architects, engineers, policymakers, building owners and occupants interested in sustainable built environments. The various successful implementations of ZEBs around the world have proven that net-zero energy performance is achievable and viable for a wide array of building types and building scales. In this process, they have established valuable principles and strategies that could be applied to a variety of upcoming ZEB projects ranging from single-family houses to large urban developments [85]. Overall, key messages from successful projects include the principle that successful ZEBs necessarily require integrated design approaches in their development.

Where conventional buildings differ is that ZEBs need to be designed in participation with all parties involved in the process: the architect, the engineer, the contractor, and the energy consultant. This would make energy efficient and the integration of renewable energy resources into intrinsic parts of the process, rather than additions. Integrated design allows synergies among various systems of buildings to be more easily identified; for instance, combining energy-efficient HVAC with passive heating and cooling strategies can achieve optimization. Another lesson learned is that the passive design principles applied are meant to reduce base energy demand. By reducing energy needs through passive design, the ZEBs can tap smaller-sized renewable energy systems that are more economical to achieve net-zero performance [85, 86]. Large-scale implementation of renewable energy systems came along with the scaling and integration experience. Solar photovoltaic panels supply most of the energy requirements in most of the ZEBs; hence, accurate energy modeling is crucial for the proper sizing of the systems [85].

All these successful initiatives have understood that substantial added value comes with diversification of the renewable energy supply, increasing the resilience of the energy supply by adding elements such as wind turbines, geothermal systems, or biomass energy. In this complex process, energy storage technologies, like battery systems or thermal storage, have been able to make significant contributions in diverse contexts with intermittent availability of renewable resources [85].

The other critical pillar for successful ZEBs is the advanced energy management systems that should be enabled through smart technologies based on artificial intelligence and machine learning. These technologies provide real-time data and make adjustments in operations based on occupancy patterns by enabling buildings to monitor, optimize, and respond dynamically to energy use. Therefore, these energy management systems permit advanced optimization and predictive maintenance [86].

The most important lesson learned in implementing ZEB involves occupant engagement and education. Successful ZEBs have invested in training programs and user-friendly interfaces for residents/users to take an active part in energy management. Also, high upfront costs remain as a major obstacle for scaling projects. Hence, making innovative financing approaches such as green bonds, public–private partnerships, adequate subsidies are required to address this obstacle. The integration of ZEB principles into retrofitting the existing stock of buildings has been more problematic compared to new buildings construction. This requires an increasingly greater need for flexible and modular design approaches to accommodate a variety of different building types [58, 87].

6.5.1 Mitigating ZEB Construction and Operation Challenges

A complex process associated with ZEB design, construction, operation face namy challenges. A significant potential is recognized for environmental and economic benefits but successful realization of ZEBs must involve solutions to several technical problems, financial barriers, regulatory limitations, and social obstacles including user behavior [88, 89]. Table 6.4 provides an overview of the primary challenges in ZEB construction and operation, showing their causes and mitigation strategies.

Table 6.4 summarizes the main challenges in ZEB construction and operation, including their causes and potential mitigation strategies. These obstacles are addressed through measures including subsidies and green bonds, targeted training, harmonized codes, performance monitoring, occupant engagement, and integrated project management. It is observed that effective challenge mitigation requires a combination of technical, financial, regulatory, and human-centered approaches.

Table 6.4 Challenges in ZEBs and corresponding mitigation measures

Challenge	Source of the challenge	Challenge mitigation
High upfront costs	ZEBs often cost more initially (materials, systems, renewables, storage) and require investment before payback	Use subsidies, long-term financing, incentives; show lifecycle cost savings; government or policy support; green bonds
Lack of technical expertise	Stakeholders (architects, contractors, operators) may not be familiar with best practices, renewable tech integration, or efficient design	Training programs; education (universities/professional development); case studies; sharing of success stories
Inconsistent or inadequate regulations	Different regions have different definitions of ZEB, different building codes, etc., which can conflict	Harmonize definitions; update codes to include ZEB standards; regulatory clarity; mandates or incentives
Monitoring, verification, performance gap	Buildings may not perform as designed; operational practices may drift; energy uses may exceed estimates	Install robust measurement and verification systems; commissioning; feedback loops; occupant engagement; digital tools
Integration of renewable energy and storage	On-site generation, storage, matching load vs generation (especially for high-rise or dense buildings) present technical/space/ cost challenges	Use hybrid systems; shared renewables; smart grid/demand management; optimizing passive design to reduce load
Behavioral issues	Occupants, building owners, contractors may resist change due to habits, lack of awareness, risk aversion	Education and outreach; incentives; demonstrating benefits; involving stakeholders early; making operations easy and user-friendly
Infrastructure limitations	Poor access to grids, lack of local renewable energy supply, or power capacity; insufficient or expensive energy storage	Coordinate with utilities; invest in grid upgrades; microgrids; local energy sources; planning for infrastructure before construction
Process fragmentation	Multiple contractors, different parties, lack of coordination; complex procurement; overlapping responsibilities	Clear project management; integrated design process; early collaboration; streamlined procurement; standards for coordination

Based on Table 6.4, one of the most significant challenges to constructing ZEBs is the relatively high initial investment required for the adoption of energy-efficient technologies and renewable energy systems. In such cases, usually when highly sophisticated renewable energy systems are involved or highly efficient insulation materials, a ZEB is more capital-intensive compared to a conventional building. Though this can be quite a costly investment for many developers, especially for those markets where construction cost is the main issue, the long-run benefits come in the form of reduced energy consumption and lower utility bills [88–90]. Complexity in the design and construction process is another issue, which requires a holistic design process in which professionals ranging from architects and engineers to energy consultants and contractors work in very close relations with each other. It is a labor-intensive process, beginning to end, that many types of professionals get involved with. In many cases, experts handling any aspect of a ZEB project must bridge the knowledge gap regarding new technologies and systems-possibly including renewable energy installations, energy storage solutions, and advanced building materials. Geographical location, climatic condition, and availability of renewable energy resources are related highly to the feasibility of the projects concerned with ZEBs [88–90]. Another important obstacle is the absence of ordinary shared standards and codes supporting ZEBs, which are still in their development in many regions. Without defined standards about the performance of a ZEB itself, it is hard to compare how projects perform and, in turn, are actually "zero energy." Ambiguity in regulations-for instance, due to unclear guidelines over energy calculation methodologies, performance testing, and certification procedures is also discouraging developers from undertaking ZEB projects [90, 91].

While significant challenges exist in the construction and operation of ZEBs, various strategies have been developed to mitigate them. As high initial costs are a major barrier, it is essential to explore innovative financing models. Government incentives include tax credits, grants, and subsidies that significantly reduce the financial burden for developers and building owners. Green bonds and energy performance contracts as financial instruments provide alternative funding opportunities that are better aligned with the long-term ZEB goals. As the volume of ZEBs built increases and their benefits become tangible, financing models may be developed to improve their financial viability [90, 91].

In fighting skill and knowledge barriers, the industry has increasingly used integrated project delivery methods that lessen the complexity of designing and constructing a ZEB. Architects, engineers, and contractors can design systems that work together because they have been working together since the beginning, which reduces inefficiencies, prevents costly redesigns, and addresses constructability issues. Increasing access to training and certification programs in the fields of professionals in sustainable design and renewable energy technologies supports closing the knowledge gap and ensures the highest standard of ZEBs that are executed [90].

To lower energy use through behavioral changes, it is necessary to combine regulations and incentives with education and awareness campaigns that promote energy-saving habits. In addition to supporting the building operators of advanced energy management systems for their best performance, proper training for effective use by the occupants and facility managers is required. Most ZEBs are equipped with clear dashboards and applications that provide real-time feedback about energy consumption, based on which users may modify their behavior. These systems can be automated and require minimal human interference to keep them working efficiently [92].

The challenge of site-specific issues involves the actualization of design solutions that consider the conditions in their locality. ZEB designers usually do highly detailed energy modeling, considering a site's specifics including solar exposure, wind pattern, and building function-quantifying how energy will be produced and consumed [92]. Advanced solar shading can reduce the need for air conditioning in hot climates, while geothermal energy systems can provide reliable heating and cooling where ground temperatures remain relatively constant over the year. Urban ZEBs could have access to solutions that are particular to limited areas, like vertical solar panels or energy storage at the community scale. Regulatory barriers, on the other hand, concern collaboration between policy makers and industry associations for detailed building codes that foster ZEBs and outline clear standards for performance. The eventual development of specific certifications to ZEB and guidelines, may standardize how projects on ZEB are carried out [91]. In addition, sustainable behavior may be further promoted, and occupant actions better aligned with the building energy goals through incentives, such as energy-saving competitions or rewards programs.

6.6 Conclusion

ZEBs represent a potential solution toward advancing sustainability in the built environment. These solution have demonstrated a potential in reducing energy consumption and mitigating carbon emissions through the application of advanced energy management systems, renewable energy technologies, and integrated design. The scalability of ZEB principles to urban developments and campus communities through shared microgrids, energy storage solutions, and smart grid integration, demonstrating the potential for community-wide sustainability. Lessons learned from global case studies show that, (to mitigate technical, financial, regulatory, and behavioral challenges) it is very important to ensure early stakeholder collaboration, integrated design approaches, occupant engagement, and innovative financing mechanisms.

References

1. Shehadi M (2020) Net-zero energy buildings: principles and applications. In: Zero-energy buildings—new approaches and technologies. IntechOpen. https://doi.org/10.5772/intechopen.92285
2. Brown D, Tokede O, Li HX, Edwards D (2024) A systematic review of barriers to implementing net zero energy buildings in Australia. J Clean Prod 467:142910. https://doi.org/10.1016/J.JCLEPRO.2024.142910
3. Mahmoodi M, Rasheed E, Le A (2024) Systematic review on the barriers and challenges of organisations in delivering new net zero emissions buildings. Buildings 14:1829. https://doi.org/10.3390/buildings14061829

4. Cao F, Su CW, Qin M, Moldovan NC (2024) The investment of renewable energy: is green bond a safe-haven to hedge U.S. monetary policy uncertainty? Energy 307. https://doi.org/10.1016/j.energy.2024.132651
5. Kristiansen AB, Ma T, Wang RZ (2019) Perspectives on industrialized transportable solar powered zero energy buildings. Renew Sustain Energy Rev 108:112–124. https://doi.org/10.1016/j.rser.2019.03.032
6. Huang B, Xing K, Ness D, Liao L, Huang K, Xie P, Huang J (2022) Rethinking carbon–neutral built environment: urban dynamics and scenario analysis. Energy Build 255:111672. https://doi.org/10.1016/J.ENBUILD.2021.111672
7. Liu Z, Sun T, Yu Y, Ke P, Deng Z, Lu C, Huo D, Ding X (2022) Near-real-time carbon emission accounting technology toward carbon neutrality. Engineering 14:44–51. https://doi.org/10.1016/J.ENG.2021.12.019
8. Durakovic B, Mujković N, Hudović-Kljuno A, Gibanica E (2025) Effect of phase change materials on the thermal behavior of biomass-based composites. Herit Sustain Dev 7:155–166. https://doi.org/10.37868/HSD.V7I1.1107
9. Martín-Chivelet N, Kapsis K, Wilson HR, Delisle V, Yang R, Olivieri L, Polo J, Eisenlohr J, Roy B, Maturi L, Otnes G, Dallapiccola M, Wijeratne WMPU (2022) Building-Integrated Photovoltaic (BIPV) products and systems: a review of energy-related behavior. Energy Build 262:111998. https://doi.org/10.1016/J.ENBUILD.2022.111998
10. Stephan A, Stephan L (2020) Achieving net zero life cycle primary energy and greenhouse gas emissions apartment buildings in a Mediterranean climate. Appl Energy 280:115932. https://doi.org/10.1016/J.APENERGY.2020.115932
11. Nord N, Tereshchenko T, Woszczek A, Næss JS, Sandberg NH, Pahlavan HM, Brattebø H (2024) A dynamic modelling approach to explore zero emission building stock opportunities towards 2050—case study of a university campus. Energy Build 325:115024. https://doi.org/10.1016/J.ENBUILD.2024.115024
12. Ivanko D, Ding Y, Nord N (2021) Analysis of heat use profiles in Norwegian educational institutions in conditions of COVID-lockdown. J Build Eng 43:102576. https://doi.org/10.1016/J.JOBE.2021.102576
13. Ang YQ, Berzolla ZM, Reinhart CF (2020) From concept to application: a review of use cases in urban building energy modeling. Appl Energy 279:115738. https://doi.org/10.1016/J.APENERGY.2020.115738
14. Ohene E, Chan APC, Darko A (2022) Review of global research advances towards net-zero emissions buildings. Energy Build 266:112142. https://doi.org/10.1016/J.ENBUILD.2022.112142
15. Aliero MS, Qureshi KN, Pasha MF, Jeon G (2021) Smart home energy management systems in Internet of Things networks for green cities demands and services. Environ Technol Innov 22:101443. https://doi.org/10.1016/J.ETI.2021.101443
16. Cabeza LF, Boquera L, Chàfer M, Vérez D (2021) Embodied energy and embodied carbon of structural building materials: worldwide progress and barriers through literature map analysis. Energy Build 231:110612. https://doi.org/10.1016/J.ENBUILD.2020.110612
17. Sharaf-Addin HH (2024) Towards net-zero carbon emissions: a systematic review of carbon sustainability reporting based on GHG protocol framework. Environ Sustain Indic 24:100516. https://doi.org/10.1016/J.INDIC.2024.100516
18. Chithambo L, Tauringana V, Tingbani I, Achiro L (2022) Stakeholder pressure and greenhouses gas voluntary disclosures. Bus Strategy Environ 31:159–172. https://doi.org/10.1002/bse.2880
19. Jin B, Bae Y (2023) Prospective research trend analysis on zero-energy building (ZEB): an artificial intelligence approach. Sustainability 15:13577. https://doi.org/10.3390/su151813577
20. Lee D, Kim J, Kim Y (2023) Il: economic evaluation of small public office buildings with class 1 of zero energy building (ZEB) in Korea by reflecting life cycle assessment (LCA). Buildings 13:1693. https://doi.org/10.3390/buildings13071693
21. Copiello S, Donati E, Bonifaci P (2024) Energy efficiency practices: a case study analysis of innovative business models in buildings. Energy Build 313:114223. https://doi.org/10.1016/J.ENBUILD.2024.114223

22. Bagaini A, Croci E, Molteni T (2022) Boosting energy home renovation through innovative business models: ONE-STOP-SHOP solutions assessment. J Clean Prod 331:129990. https://doi.org/10.1016/J.JCLEPRO.2021.129990
23. Seto KC, Churkina G, Hsu A, Keller M, Newman PWG, Qin B, Ramaswami A (2021) From low- to net-zero carbon cities: the next global agenda. Annu Rev Environ Resour 46:377–415. https://doi.org/10.1146/annurev-environ-050120-113117
24. Ensign P, Roy S, Brzustowski T (2021) Decisions by key office building stakeholders to build or retrofit green in Toronto's urban core. Sustainability 13:6969. https://doi.org/10.3390/su13126969
25. Guo K, Zhang L (2020) Guarantee optimization in energy performance contracting with real option analysis. J Clean Prod 258:120908. https://doi.org/10.1016/j.jclepro.2020.120908
26. Bullitt Center. https://bullittcenter.org/. Last Accessed 14 Apr 2025
27. Yang T, Liu W, Kramer GJ, Sun Q (2021) Seasonal thermal energy storage: a techno-economic literature review. Renew Sustain Energy Rev 139:110732. https://doi.org/10.1016/J.RSER.2021.110732
28. Tian Z, Zhang S, Deng J, Fan J, Huang J, Kong W, Perers B, Furbo S (2019) Large-scale solar district heating plants in Danish smart thermal grid: developments and recent trends. Energy Convers Manag 189:67–80. https://doi.org/10.1016/J.ENCONMAN.2019.03.071
29. Chen S, Zhu Y, Chen Y, Liu W (2020) Usage strategy of phase change materials in plastic greenhouses, in hot summer and cold winter climate. Appl Energy 277:115416. https://doi.org/10.1016/J.APENERGY.2020.115416
30. Han F, Chen C, Hu Q, He Y, Wei S, Li C (2021) Modeling method of an active–passive ventilation wall with latent heat storage for evaluating its thermal properties in the solar greenhouse. Energy Build 238:110840. https://doi.org/10.1016/J.ENBUILD.2021.110840
31. Alawode A, Rajagopalan P (2022) Feasibility of net zero energy high rise apartment buildings in Australia. Sol Energy 231:158–174. https://doi.org/10.1016/J.SOLENER.2021.11.030
32. Oh S, Gardner JF (2022) Energy consumption analysis using measured data from a net-zero energy commercial building in a cold and dry climate. Sustainability 14:10346. https://doi.org/10.3390/SU141610346
33. Khan HS, Asif M, Mohammed MA (2017) Case study of a nearly zero energy building in Italian climatic conditions. Infrastructures 2:19. https://doi.org/10.3390/INFRASTRUCTURES2040019
34. Farizal F, Noviandri MA, Hamdani H (2024) Sustainability development through a nearly zero energy building implementation case: an office building in South Jakarta. Sustainability 16:7027. https://doi.org/10.3390/SU16167027
35. Ismail RF, Elrazaz ZM, Abdelsalam NF (2020) An intensive methodology for designing near zero energy public buildings in egypt, office building as a case study. JES J Eng Sci 48:447–467. https://doi.org/10.21608/JESAUN.2020.108951
36. Ascanio-Villabona J, Orejarena-Osorio N, Castillo-Leon NY, Tarazona-Romero BE, Jaimes-Quintero KT (2024) Thermal comfort evaluation using different glazing technologies in an energy model of rural housing in a cold climate. Herit Sustain Dev 6:735–748. https://doi.org/10.37868/HSD.V6I2.922
37. Tarazona-Romero BE, Socha-Rojas ND, Ascanio-Villabona J, Kafarov V, Castillo-Leon NY (2024) Performance evaluation of solar thermal collectors in Colombian thermal floors by dynamic simulation. Sustain Eng Innov 6:421–432. https://doi.org/10.37868/SEI.V6I2.ID394
38. Pascal M, Margolis HG, Etzel RA (2021) Greening the international society for environmental epidemiology. Epidemiology 32:466–468. https://doi.org/10.1097/EDE.0000000000001355
39. Shi F, Wang S, Huang J, Hong X (2020) Design strategies and energy performance of a net-zero energy house based on natural philosophy. J Asian Arch Build Eng 19:1–15. https://doi.org/10.1080/13467581.2019.1696206
40. Feng W, Zhang Q, Ji H, Wang R, Zhou N, Ye Q, Hao B, Li Y, Luo D, Lau SSY (2019) A review of net zero energy buildings in hot and humid climates: experience learned from 34 case study buildings. Renew Sustain Energy Rev 114:109303. https://doi.org/10.1016/J.RSER.2019.109303

41. Rey-Hernández JM, Velasco-Gómez E, San José-Alonso JF, Tejero-González A, Rey-Martínez FJ (2018) Energy analysis at a near zero energy building. A case-study in Spain. Energies 11:857. https://doi.org/10.3390/EN11040857
42. Ke Z, Liu X, Zhang H, Jia X, Zeng W, Yan J, Hu H, Hien WN (2023) Energy consumption and carbon emissions of nearly zero-energy buildings in hot summer and cold winter zones of China. Sustainability 15:11453. https://doi.org/10.3390/SU151411453
43. Sedat V, Sercan İ, İlhan S, Emre Ü. Case study on improving the energy performance of existing buildings towards net zero or nearly zero energy building transition. https://doi.org/10.5281/ZENODO.13117697
44. Senturk A, Ozcan M (2024) Nearly zero energy building design and optimization: a residential building transformation in Türkiye. Energy Explor Exploit 42:997–1026. https://doi.org/10.1177/01445987231219765
45. Preghenella N, Battistella C (2021) Exploring business models for sustainability: a bibliographic investigation of the literature and future research directions. Bus Strategy Environ 30:2505–2522. https://doi.org/10.1002/bse.2760
46. Zhao X, Pan W (2022) The characteristics and evolution of business model for green buildings: a bibliometric approach. Eng Constr Archit Manag 29:4241–4266. https://doi.org/10.1108/ECAM-08-2020-0657
47. Comin LC, Aguiar CC, Sehnem S, Yusliza M-Y, Cazella CF, Julkovski DJ (2019) Sustainable business models: a literature review. Benchmarking Int J. 27:2028–2047. https://doi.org/10.1108/BIJ-12-2018-0384
48. Harputlugil T, de Wilde P (2021) The interaction between humans and buildings for energy efficiency: a critical review. Energy Res Soc Sci 71:101828. https://doi.org/10.1016/J.ERSS.2020.101828
49. Anarene B (2024) Revolutionizing energy efficiency in commercial and institutional buildings: a complete analysis. Int J Sci Res Manag (IJSRM) 12:7444–7468. https://doi.org/10.18535/IJSRM/V12I09.EM12
50. Vyas S, Sharma A (2023) Energy efficiency in commercial buildings: challenges, opportunities, and solutions (review). J Energy Conserv 1:7–10. https://doi.org/10.14302/ISSN.2642-3146.JEC-23-4568
51. Hongvityakorn B, Jaruwasupant N, Khongphinitbunjong K, Aggarangsi P (2024) Achieving nearly zero-energy buildings through renewable energy production-storage optimization. Energies 17:4845. https://doi.org/10.3390/EN17194845
52. Lou HL, Hsieh SH (2024) Towards zero: a review on strategies in achieving net-zero-energy and net-zero-carbon buildings. Sustainability (Switzerland) 16:4735. https://doi.org/10.3390/SU16114735/S1
53. Zaferanchi M, Sozer H (2024) Effectiveness of interventions to convert the energy consumption of an educational building to zero energy. Int J Build Pathol Adap 42:485–509. https://doi.org/10.1108/IJBPA-08-2021-0114/FULL/XML
54. Parvin K, Hossain MJ, Arsad AZ, Ker PJ, Hannan MA (2025) Building energy technologies towards achieving net-zero pathway: a comprehensive review, challenges and future directions. J Build Eng 100:111795. https://doi.org/10.1016/J.JOBE.2025.111795
55. Balasbaneh AT, Lye TY (2024) Investigation of barriers and improvement methods for net zero energy buildings: stakeholders' viewpoints. J Electr Syst 20:6267–6273
56. Falana J, Osei-Kyei R, Tam VW (2024) Towards achieving a net zero carbon building: a review of key stakeholders and their roles in net zero carbon building whole life cycle. J Build Eng 82:108223. https://doi.org/10.1016/J.JOBE.2023.108223
57. Nidhin BKSN, Domingo N, Bui TTP, Wilkinson S (2025) Construction stakeholders' knowledge on zero carbon initiatives in New Zealand. Int J Build Pathol Adap 43:512–527. https://doi.org/10.1108/IJBPA-08-2022-0119
58. Souaid C, van der Heijden H, Elsinga M (2021) Institutional barriers to near zero-energy housing: a context specific approach. Sustainability 13:7135. https://doi.org/10.3390/SU13137135

59. Cárcel-Carrasco J, Peñalvo-López E, Morcillo I (2017) Legal challenges in the implementation of Nearly Zero Energy Buildings (nZEB) in Spain. VITRUVIO-Int J Arch Technol Sustain 2:1–11. https://doi.org/10.4995/VITRUVIO-IJATS.2017.7404
60. Briga-Sá A, Cristina E, Luso P, Manuel P, Brandão I, Kalumba S, Volker H, Nurick S (2025) An investigation of the adoption of Net-Zero Buildings (NZBs) in the South African commercial property market. Sustainability 17:5272. https://doi.org/10.3390/SU17125272
61. Faulconbridge J, Cass N, Connaughton J (2018) How market standards affect building design: the case of low energy design in commercial offices. Environ Plan A Econ Space 50:627–650. https://doi.org/10.1177/0308518X17752681
62. Maduta C, D'Agostino D, Tsemekidi-Tzeiranaki S, Castellazzi L (2025) From nearly zero-energy buildings (NZEBs) to zero-emission buildings (ZEBs): current status and future perspectives. Energy Build 328:115133. https://doi.org/10.1016/J.ENBUILD.2024.115133
63. Akomea-Frimpong I, Kukah AS, Jin X, Osei-Kyei R, Pariafsai F (2022) Green finance for green buildings: a systematic review and conceptual foundation. J Clean Prod 356:131869. https://doi.org/10.1016/J.JCLEPRO.2022.131869
64. Attia S, Eleftheriou P, Xeni F, Morlot R, Ménézo C, Kostopoulos V, Betsi M, Kalaitzoglou I, Pagliano L, Cellura M, Almeida M, Ferreira M, Baracu T, Badescu V, Crutescu R, Hidalgo-Betanzos JM (2017) Overview and future challenges of nearly zero energy buildings (nZEB) design in Southern Europe. Energy Build 155:439–458. https://doi.org/10.1016/J.ENBUILD.2017.09.043
65. Tejani A, Yadav J, Toshniwal V, Gajjar H. Achieving net-zero energy buildings: the strategic role of HVAC systems in design and implementation. 2:4. https://doi.org/10.56472/25832646/JETA-V2I1P107
66. Dimitriou S, Kyprianou I, Papanicolas CN, Serghides D (2020) A new approach in the refurbishment of the office buildings—from standard to alternative nearly zero energy buildings. Int J Sustain Energ 39:761–778. https://doi.org/10.1080/14786451.2020.1749629
67. Lam KH, To WM, Lee PKC (2022) Smart building management system (SBMS) for commercial buildings—key attributes and usage intentions from building professionals' perspective. Sustainability 15:80. https://doi.org/10.3390/SU15010080
68. Kang JE, Ahn KU, Park CS, Schuetze T, Rosen MA, Tieben H, Chelleri L, Ostermeyer Y, Wolfram M (2015) Assessment of passive vs. active strategies for a school building design. Sustainability 15136–15151. https://doi.org/10.3390/SU71115136
69. Mohammed GA, Mabrouk M, He G, Abdrabo KI (2023) Towards sustainable cities: a review of zero energy buildings techniques and global activities in residential buildings. Energies 16:3775. https://doi.org/10.3390/EN16093775
70. Utaminingsih A, Widowati SY, Witjaksono EH (2024) Sustainable business model innovation: external and internal factors on SMEs. Int J Innov Sci 16:95–113. https://doi.org/10.1108/IJIS-04-2022-0061
71. Yevu SK, Owusu EK, Chan APC, Oti-Sarpong K, Wuni IY, Tetteh MO (2023) Systematic review on the integration of building information modelling and prefabrication construction for low-carbon building delivery. Build Res Inf 51:279–300. https://doi.org/10.1080/09613218.2022.2131504
72. Liu J, Chen X, Yang H, Shan K (2021) Hybrid renewable energy applications in zero-energy buildings and communities integrating battery and hydrogen vehicle storage. Appl Energy 290:116733. https://doi.org/10.1016/J.APENERGY.2021.116733
73. Lindholm O, Rehman HU, Reda F (2021) Positioning positive energy districts in European cities. Buildings 11:19. https://doi.org/10.3390/BUILDINGS11010019
74. Wang WT, Yang H, Xiang CY (2023) Green roofs and facades with integrated photovoltaic system for zero energy eco-friendly building—a review. Sustainable Energy Technol Assess 60:103426. https://doi.org/10.1016/J.SETA.2023.103426
75. Wijewansha AS, Tennakoon GA, Waidyasekara KGAS, Ekanayake BJ (2021) Implementation of circular economy principles during pre-construction stage: the case of Sri Lanka. Built Environ Proj Asset Manag 11:750–766. https://doi.org/10.1108/BEPAM-04-2020-0072

76. Garcia YV, Garzon O, Andrade F, Irizarry A, Rodriguez-Martinez OF (2022) Methodology to implement a microgrid in a university campus. Appl Sci 12. https://doi.org/10.3390/APP12094563
77. Fiorentzis K, Katsigiannis Y, Karapidakis E, Versaci M, Paspatis A, Fiorentzis K, Katsigiannis Y, Karapidakis E (2022) Smart campus microgrids towards a sustainable energy transition—the case study of the Hellenic Mediterranean university in Crete. Mathematics 10:1065. https://doi.org/10.3390/MATH10071065
78. Duraković B (2020) PCMs in separate heat storage modules. In: PCM-based building envelope systems: innovative energy solutions for passive design. Springer, pp 121–146. https://doi.org/10.1007/978-3-030-38335-0_6
79. Huylo M, Taheri S, Novoselac A (2025) Integration of renewable energy generation and storage systems for emissions reduction in an islanded campus microgrid. Build Environ 274. https://doi.org/10.1016/J.BUILDENV.2025.112736
80. Rana MJ, Hasan MR, Sobuz MHR, Sutan NM (2021) Evaluation of passive design strategies to achieve NZEB in the corporate facilities: the context of Bangladeshi subtropical monsoon climate. Int J Build Pathol Adap 39:619–654. https://doi.org/10.1108/IJBPA-05-2020-0037
81. UC Davis West Village, https://westvillage.ucdavis.edu/?utm_source=chatgpt.com. Last Accessed 6 Oct 2025
82. Nanyang Technological University. https://www.tropicalenvironment.com.sg/nanyang-technological-university?utm_source=chatgpt.com. Last Accessed 6 Oct 2025
83. Building a Sustainable Future with Zero Energy Buildings—CTC. https://www.ctc.edu.za/civil-engineering/building-a-sustainable-future-with-zero-energy-buildings/?utm_source=chatgpt.com. Last Accessed 6 Oct 2025
84. Zero-Energy Buildings: Sustainable Architecture, https://businesscasestudies.co.uk/what-are-zero-energy-buildings/?utm_source=chatgpt.com. Last Accessed 6 Oct 2025
85. Vares S, Häkkinen T, Ketomäki J, Shemeikka J, Jung N (2019) Impact of renewable energy technologies on the embodied and operational GHG emissions of a nearly zero energy building. J Build Eng 22:439–450. https://doi.org/10.1016/J.JOBE.2018.12.017
86. Moschetti R, Brattebø H, Sparrevik M (2019) Exploring the pathway from zero-energy to zero-emission building solutions: a case study of a Norwegian office building. Energy Build 188–189:84–97. https://doi.org/10.1016/j.enbuild.2019.01.047
87. Zhao J, Abdul Aziz F, Deng Y, Ujang N, Xiao Y (2024) A review of comprehensive post-occupancy evaluation feedback on occupant-centric thermal comfort and building energy efficiency. Buildings 14:2892. https://doi.org/10.3390/BUILDINGS14092892/S1
88. Myint NN, Shafique M, Zhou X, Zheng Z (2025) Net zero carbon buildings: a review on recent advances, knowledge gaps and research directions. Case Stud Constr Mater e04200. https://doi.org/10.1016/j.cscm.2024.e04200
89. Godin K, Sapinski JP, Dupuis S (2021) The transition to net zero energy (NZE) housing: an integrated approach to market, state, and other barriers. Clean Responsible Consum 3:100043. https://doi.org/10.1016/j.clrc.2021.100043
90. Pan W, Pan M (2021) Drivers, barriers and strategies for zero carbon buildings in high-rise high-density cities. Energy Build 242:110970. https://doi.org/10.1016/j.enbuild.2021.110970
91. Bui TTP, Domingo N, MacGregor C, Wilkinson S (2022) Zero carbon refurbishment for existing buildings: a literature review. IOP Conf Ser Earth Environ Sci 1101:22017. https://doi.org/10.1088/1755-1315/1101/2/022017
92. Serrano-Jiménez A, Femenías P, Thuvander L, Barrios-Padura Á (2021) A multi-criteria decision support method towards selecting feasible and sustainable housing renovation strategies. J Clean Prod 278:123588. https://doi.org/10.1016/j.jclepro.2020.123588

Chapter 7
From Zero-Energy Buildings to Zero Energy Communities

Abbreviations

GHG	Greenhouse gas
NZECs	Net-Zero Energy Communities
RES	Renewable Energy Sources
ZEBs	Zero Energy Buildings
ZEC	Zero Energy Communities

7.1 Introduction

The growing demand for environmental sustainability has sparked the search for innovative solutions like Net-Zero Energy Communities (NZECs), which move beyond the scope of Zero-Energy Buildings (ZEBs). While ZEBs focus on individual buildings, NZECs extend the concept of ZEBs to a larger scale, typically a neighborhood or a community incorporating not just individual buildings but also shared resources. These communities aim to balance energy consumption with renewable energy generation within or near that community [1]. Net-zero energy communities or else known as Zero Energy Communities (ZEC) or NZEC aim to significantly reduce energy through gaining efficiency seen by balancing the energy used in everyday life such as thermal, vehicle, and electrical energy that's in a community creating more renewable energy [2]. The Click or tap here to enter text. NZEC is comprised by four perspectives according to the U.S. Department of Energy [3]:

- A central energy system that should include renewable energy sources and can accommodate energy for the whole community [3].
- The percentage of energy loss should be kept in mind to deliver energy [3].
- Financial aspects of the community should be considered [3].

B. Duraković, *Advancing Zero Energy Buildings*, Green Energy and Technology,
https://doi.org/10.1007/978-3-032-16601-2_7

- Environmental impacts including greenhouse gas (GHG) emissions are to be considered [3].

This chapter presents case studies that analyze how policy mixes and renewable energy technologies contribute to the development of net-zero communities. These technologies, such as solar panel systems, wind power, bioenergy, and hydropower, demonstrate significant potential to create positive environmental impacts.

7.2 Status of Renewable Energy on a Global Scale

7.2.1 The Importance of Net-Zero Energy Communities

The use of fossil fuels and other means of unsustainability lead to the main issues of global warming and climate change, strengthening the need to aim for green energy and sustainable development [4]. Looking at the Paris agreement, it states that the global average temperature should be aimed at less than 1.5 °C, and all global parties should contribute to reducing national emissions by creating actions and targets [5].

To achieve these targets, the transition from conventional energy systems to low-carbon and renewable-based frameworks has become a global priority. In this context, NZECs emerge as practical and scalable solutions, integrating energy efficiency, renewable generation, and smart management technologies. They demonstrate how local-level action can contribute meaningfully to global goals, transforming urban and rural spaces into self-sufficient energy ecosystems.

There are three major sectors which contribute to the wide range of energy consumption including industry, transportation, and building sector where building takes up to 40% of world's total energy use [3].

Given the high share of energy use in buildings, decarbonizing this sector represents one of the most impactful strategies for emission reduction. NZECs play a central role here, as they combine sustainable building practices with renewable integration, reducing both operational and embodied carbon footprints.

If adopting renewable energy, statistics show that a 1% increase in the use of renewable energy can lead to a 1.4% decrease in CO_2 emissions [6]. The Net-Zero Communities aim to create a balance of energy regarding input and output providing immense opportunities economically by technological advancements, and 'green jobs' that positively impact the environment by adapting this within communities. Table 7.1 presents policy mixes for various transition functions for selected countries towards achieving the Paris Agreement goals.

The comparison in Table 7.1 illustrates how leading economies are shaping their policy environments to foster innovation and clean energy adoption. While each region employs a unique mix of incentives and regulations, the overall trajectory reflects a shared commitment to decarbonization and sustainable growth. This convergence of policy strategies forms the foundation for scaling NZECs globally.

Table 7.1 Policy mixes and transition functions in China, the EU and the USA 1

Policy area	Purpose	China	EU	USA	Ref
Supply-push policies	Fund public research and innovation to create and lower costs of low-carbon tech	Expanding efforts with good coordination	Leading globally but coordination needs improvement	Leading with strong coordination	[7, 8]
Demand-pull policies	Encourage investment and market uptake of clean technologies	Gradual, steady policy improvements	Ambitious targets and strengthening policies	Federal climate policies fragmented but renewable policies durable	[8, 9]
Green industrial policies	Foster green jobs and reduce support for polluting industries	Both polluting activities and green jobs are growing	Active policies to replace polluting sectors with green alternatives	Market-driven shift from polluting to green industries	[9, 10]
Distributional policies (energy justice)	Ensure no groups are left behind, supporting public acceptance	Coal remains central due to economic and poverty concerns	Funds allocated for just transition	Emerging focus on just transition funding	[7, 11]
Policy implementation and reform	Use domestic experience to progressively strengthen climate policy	Challenges exist but some local governments lead	Reforms informed by experience; some countries lead the way	Political divisions hinder progress; some states act as pioneers	[12, 13]
Economic shocks and recovery packages	Leverage crises to boost green growth	Some use of crisis to promote green investments	Crises mainly used to reinforce climate policies	Democrats advocate linking recovery to green growth	[11, 14]

These policies discussed in Table 7.1 above display the transitions needed to create "green growth" [14]. On a global scale, the way to transition to NZEC is through innovative technologies which create an important factor in applying these changes. The EU adopted a decision-making strategy which looks to act strategically with climate policy whilst thinking about the long term [14].

Collaboration between governments, research institutions, and private stakeholders remains vital for accelerating NZEC implementation. Access to research and innovation funding, along with knowledge-sharing initiatives, helps overcome technical and financial barriers that often limit the transition to low-carbon communities.

Beyond technological and environmental performance, the transition to Net-Zero Energy Communities also depends on social acceptance, participation, and shared ownership of sustainability goals. Successful NZECs often integrate community engagement strategies, promoting collaboration among residents, local authorities, and developers. This ensures that energy-efficient interventions are not only implemented but maintained over time, and that communities develop resilience and social cohesion alongside environmental sustainability [15, 16]. This way, NZECs can foster both ecological and social well-being, turning energy transitions into a communal endeavor.

The development of NZECs is also shaped by broader policy frameworks that guide investment, innovation, and urban planning. In Europe, the European Green Deal (2019) and the EU Renovation Wave Strategy (2020) aim to achieve climate neutrality by 2050, emphasizing the modernization of building stocks and the integration of renewable energy at the community level. These initiatives highlight the importance of local-scale energy transitions, demonstrating how national and regional policy alignment can accelerate the adoption of net-zero strategies while providing financial and technical support to communities [17, 18].

To create advancements for a sustainable future, global efforts are made by adapting public 'push' policies transitioning to low-carbon landscapes [14]. Main key strategies for these global efforts are shifting towards cleaner energy sources, promoting energy efficiency, and facilitation of the Carbon Capture and Storage solution [14]. Initiated in 2015 by 22 nations, the mission of innovation is to double R&D funding for low-carbon energy by 2021, global budgets then have risen with China increasing by 10% and USA and EU by 7% [14]. The funding opportunities coming from global initiatives like Mission Innovation can provide NZECs with opportunities for collaboration and funding partnerships. The access to these fundings can further support the development and implementation of innovative solutions to create NZECs.

Building on these lessons from policy, community engagement, and post-conflict reconstruction, the following section examines the growth of renewable energy technologies as the foundation for Net-Zero Energy Communities.

7.2.2 Growth of Renewable Energy

The use of renewable energy has notably grown in the past two decades as popularity of sustainable energy and its movement is increasingly becoming cost competitive against fossil fuels [15, 16]. According to the EU commission, renewable energy sources are energy sources that nature renews at a faster rate than they are used. That includes renewable non-fossil energy sources such as solar, wind, wave, tidal, hydropower, biomass, landfill gas, geothermal, sewage treatment plant gas and biogases [19].

Solar energy is the source of energy that is received from the Sun in form of radiation. In a day it produces more energy than the current population can produce in

27 years. It is free and inexhaustible source which can be used instead of fossil fuels [19]. Most commonly, it is harvested using Solar Photovoltaics (PV), technology that enables users to convert energy from the sun into light or electricity through photovoltaic panels [20]. The use of solar energy is still quite low in terms of global statistics as it contributes 3.6% to global energy production. However, compared to other renewable energy technologies, it shows that 31% of PV was installed in 2022, positioning second after hydropower energy installed [21]. As approximately 85% of energy used in households is allocated to heating and hot water, adapting solar energy in heating strategies can lead to increasing efficiency and reducing GHG emissions of the surrounding environment [22].

To maximize the effectiveness of renewable energy in Net-Zero Energy Communities, energy storage systems and smart grid technologies are essential. Storage solutions, such as batteries or thermal storage, help balance fluctuations in solar or wind generation and ensure a reliable supply of electricity and heat. Smart grids enable real-time monitoring and adaptive management of energy flows, allowing communities to optimize consumption, integrate distributed energy sources, and reduce reliance on fossil fuels [23, 95]. These technologies make it possible for NZECs to maintain a stable energy balance and fully realize the potential of renewable energy sources.

Wind energy uses the natural movement of air to power wind turbines, which ultimately generate electricity [24]. The outcome of the wind power usually depends on wind speed. The potential outcome of wind energy globally is quite large with a higher technical potential of generating electricity seen onshore with the estimated result of generating 20,000–50,000 terawatt hours per year (THh/y) [25].

Geothermal energy is another renewable source which can be applied to NZEC. This energy is taken from the heat of the Earth, which is sustainable and clean. It can range from low grounds such as hot water or rocks placed a few kilometers below the earth surface, or deeper from Earth molten rock or magma which is very high in temperatures [26]. Harvested from these sources, geothermal energy has immense potential for reducing our need to use oil and further creating less harm to the environment [27, 28].

Biomass and Waste-to-Energy Relevance

In addition to solar, wind, and geothermal, biomass and waste-to-energy systems provide complementary renewable solutions for NZECs. Organic waste, agricultural residues, or dedicated energy crops can be converted into heat, electricity, or biogas, offering local energy generation while reducing environmental impact. These systems also contribute to circular economy goals by turning waste into a resource and can be particularly beneficial in communities with limited access to conventional fuels [29, 30].

Adoption of renewable energy technologies varies globally due to differences in policy support, infrastructure, and investment capacity. Regions with strong governmental incentives, regulatory frameworks, and public funding, such as the European Union, have achieved faster deployment of community-scale renewable solutions. This demonstrates that technological availability alone is insufficient; coordinated policy measures, financial incentives, and public engagement are essential to accelerate the transition to NZECs [17, 31].

In summary, the global growth of renewable energy technologies and the policy frameworks supporting them form the foundation for Net-Zero Energy Communities. The convergence of international climate goals, national innovation programs, and local energy strategies demonstrates that decarbonization is achievable when communities act as active participants in the energy transition. The following section examines how these global dynamics translate into local implementation, emphasizing the spatial and social dimensions of energy transition.

7.3 Government Policy Framework

Global climate initiatives, such as the Paris Agreement and the European Green Deal, provide the overarching framework within which national strategies like the UK's Net Zero Strategy operate. These agreements encourage countries to set emission reduction targets, support research and innovation in low-carbon technologies, and facilitate international collaboration. Programs like Mission Innovation, launched in 2015, aim to double public R&D investments in clean energy technologies, offering opportunities for governments to enable the development and deployment of Net-Zero Energy Communities [18, 32]. By aligning national policies with international objectives, governments create a cohesive pathway for achieving sustainable energy transitions at multiple scales.

The research and development (R&D) programs have put an emphasis on the progression of renewable technologies. Besides technological advancements, government policies have played a significant role in promoting the widespread adoption of renewable energy. These policies are a foundation for communities to prioritize sustainability and create reliable energy supply across national, regional, and local areas [33].

Popular policies can be evident through tax incentives and subsidies which have been taking place globally, growing further interest in cleaner energy development [33].

In addition to fiscal incentives, regulatory instruments constitute a critical dimension of government policy frameworks supporting Net-Zero Energy Communities. For example, building energy codes that mandate "zero-carbon-ready" performance for new constructions by 2030 are identified by the International Energy Agency (IEA) as among the most cost-effective means of reducing emissions in the built environment [34, 35]. Alongside codes, policy packages now increasingly integrate disclosure programs, mandatory energy performance certificates, and compliance-path regulations for existing buildings undergoing renovation [36, 37].

When these regulatory instruments are combined with incentive-based approaches, such as subsidies, tax credits, and grants, a comprehensive policy mix emerges that supports both the *supply-push* (innovation) and *demand-pull* (market adoption) sides of the energy transition. This integrated regulatory-incentive strategy brings structural change to energy systems and underpins the formation of sustainable, community-scale energy ecosystems.

The UK is a prime example of creating policy frameworks to establish net-zero communities, as they have already in the last three decades reduced their emissions by 44% whilst growing the economy by over 75%. Their way to net-zero is not through fast processes; it is rather implemented slowly which will have a better longer lasting outcome [38]. The Net Zero Strategy outlines policies and ideas which would reduce carbon emissions across all the UK and its economy, whilst achieving net-zero across country wide by 2050 [39]. The "Ten Point Plan for a Green Industrial Revolution" also published by the UK government sets out paths to accelerate to net-zero [40]. This includes prioritization of the offshore wind, driving the growth of low carbon hydrogen, delivering advanced nuclear power, accelerating the shift to zero emission vehicles, green public transport, jet zero and green ships, greener buildings, investing in carbon capture, protecting our natural environment, and green finance and innovation [40, 41]. The 2022 progress in reducing emissions in the UK state that in 2021 UK's GHG was 447 million metric tons, approximately 47% lower than in 1990, a 10% drop from 2019 [42].

To reduce emissions and reach the overall goal of net zero by 2050, more than half of the emissions should be reduced by citizens making eco-friendly choices [43]. Specific changes are focused on transportation, heating within homes and food [43]. For some time now, there has been a willingness and effort within local communities to try using renewable energy sources such as the adaptation of solar powered energy which is quite common and popular. However, this is usually on a small scale and does not contribute to a larger market that can create a significant impact [43, 44].

Effective NZEC policies also rely on active community engagement and public participation. Local governments, planners, and NGOs implement educational programs, workshops, and incentive schemes to raise awareness of energy efficiency, renewable energy technologies, and sustainable practices. By fostering citizen participation in energy planning and decision-making, communities can collectively adopt low-carbon behaviors, implement micro-generation projects, and meaningfully contribute to national net-zero targets [45] [46]. Integrating social and behavioral strategies ensures that policy implementation extends beyond mere technology adoption, creating a culture of sustainability and enhancing the long-term resilience of energy systems.

Further policies that are applied also establish growth for employment in low carbon businesses, estimating an increase in 16% [47]. By implementing proper net zero policy frameworks, this would lead to hundreds of billions of private investments by 2030. Through the development of low-carbon industries and technologies, there would be a significant increase in new jobs as well, boosting the economy [48].

A policy which has been implemented in the UK is the "Clean Air Strategy". The World Health Organization mentioned this as "an example for the rest of the world to follow". It tackles the concentration of nitrogen dioxide or 'The NO_2 plan' which is to achieve better quality in areas breaching legal limits [49]. This adapted policy seen all over the UK states that if the vehicle of an individual exceeds emissions standards, then a tax is paid if driving through areas that are under law and are in clean air zones. This further leads to a reduction of emissions in areas that are already in high danger of air pollution and to motivate citizens to purchase low emission vehicles [50]. Therefore, these policies and frameworks that are applied lead to the growth of NZEC, as these drive not just one but many towards a sustainable and low emission standard.

To ensure the effectiveness of net-zero energy community policies, governments must establish robust monitoring, reporting, and evaluation (MRE) frameworks that systematically track energy consumption patterns, greenhouse-gas emissions, and stakeholder engagement. Such frameworks serve as critical feedback loops, enabling timely adjustments to strategies, highlighting implementation gaps, and reinforcing accountability [51]. For example, the European Commission's "Enabling framework for renewables" highlights how streamlined permitting, transparent data collection and annual reporting improve the governance of community-scale renewable projects [52]. By embedding these mechanisms into policy design, governments can foster adaptive, evidence-driven transitions, scale successful net-zero initiatives and linking technical outcomes with social and economic indicators.

7.4 Insights from Net-Zero Community Initiatives

Examining successful net-zero community initiatives provides practical insights into the integration of renewable energy technologies, occupant behavior, and policy frameworks at the community scale.

Building on the preceding discussion of government policies and institutional frameworks, it becomes evident that translating national and international energy strategies into tangible outcomes requires localized, community-based implementation. NZECs serve as the operational extension of these policy ambitions, demonstrating how strategic planning, citizen participation, and technological integration can collectively deliver measurable progress toward decarbonization. Examining real-world NZEC initiatives therefore provides critical insights into how policy intent is materialized at the spatial and social levels.

In academic literature, net-zero communities are recognized as promising examples that utilize renewable energy and energy-efficient technologies to significantly reduce carbon emissions. Success stories encourage others to replicate these strategies in their own communities. People, as the main source for community engagement and energy, must be able to willingly participate in energy efficiency and sustainable initiatives [53]. Despite improvements in building energy efficiency, overall energy consumption is highly dependent on occupant behavior [54].

A reliable supply of electricity is the main driver of sustained economic growth and social development. However, still approximately 1.1 billion people worldwide live without access to electricity [55]. Mostly people who live in rural or underdeveloped countries are out of reach [56]. For example, in Pakistan, about 85% of residents in rural areas do not have electricity as they live remote and far from each other [57]. Trying to connect these lines is not profitable, as people within rural areas need only 50–100 watts for each home, making it less attractive to provide electrical services [58]. Therefore, the approach for Photovoltaic installations could be a permissible solution for off grid areas as these provide rural electrification and eliminate shipment fuel charges with onsite use of renewable sources [59, 60]. A positive example could be Buriro, Pakistan, a small village of 800 inhabitants whose primary economy is based on farming. Its climate is quite hot with temperatures ranging from 38 to 47 degrees Celsius [58]. To develop the NZEC model, an estimation of various loads was surveyed to analyze energy consumption and chose the main system for the rural area [58].

A software tool, HOMER PROGRAM or hybrid optimization for electric renewables, which is designed for both off-grid and grid connected power systems [61], was deployed to model and optimize power systems using renewable sources, storage, and backup generators which minimizes the cost whilst meeting energy needs [62]. This technique was used to simulate what can be done in the area to create better energy efficient sources. Since Buriro has a desert climate, it receives high solar irradiance, creating 6.16 kWh/m^2/day, with a high potential for solar energy generation [58].

The study in the city of Alexandroupolis, Greece, investigates the feasibility of establishing both a newly built positive-energy neighborhood and a near-zero-energy retrofit district as part of the EU's transition to renewable- and smart-energy communities. It evaluates combinations of high-performance building envelopes (insulation beyond current regulations), on-site renewable energy generation (PV systems), energy storage, ground-source heat pumps, and district heating/cooling networks, applying a parametric techno-economic analysis at district scale. Results show that with appropriate technology packages, communities can meet near-zero or net-zero energy targets, and significantly reduce greenhouse-gas emission, while demonstrate that applying retrofit and generation measures at the community level is more cost-effective than treating buildings individually. The paper offers a roadmap for scaling smart, low-carbon energy communities and underscores the importance of district-scale planning, local renewable self-supply, and enabling policy to drive the shift from conventional fossil-based systems toward sustainable energy-smart neighborhoods [63].

Building on the Greek near-zero/positive energy community example, the study by Modelling renewable energy communities: assessing the impact of different configurations, technologies and types of participants examines a real-world case in the parish of Areeiro, Lisbon, modelling various configurations of a renewable energy community (EC) with diverse participant types (residential consumers and prosumers, industry, university, retail), technologies (PV solar arrays, battery storage, electric vehicles) and local energy market models (collective self-consumption vs. peer-to-peer trading). The findings show electricity cost savings for communities of up to 42%, and individual prosumer savings of 48–53%, while community self-sufficiency rates reached up to 12.5% under a high-solar-fraction scenario. The study underscores that to maximize GHG reductions and economic benefit, ECs need larger PV deployments, higher prosumer adoption rates and well-designed market/regulatory frameworks, making this a compelling example of how smart energy communities can transition from fossil-based supply to decentralized renewable systems [64].

Building on the Greek experience of near-zero and positive-energy communities, the study by Net Zero Energy Communities: Integrated Power System, Building and Transport Sectors expands the framework for community-scale sustainability by emphasizing the need to integrate the power system, buildings and transportation (PBT) sectors. The authors argue that many current "net-zero community" definitions remain limited to buildings alone, while neglecting the broader energy use and GHG emissions associated with mobility and grid infrastructure. They review multiple planned NZC cases globally, identify inconsistent metrics and unclear system boundaries as key obstacles, and propose a design roadmap that prioritizes early planning, high-performance envelopes, electrification of end-uses and maximized on-site renewables. This integrated PBT-centric approach exemplifies how whole-community systems thinking can support the transition from fossil-based urban systems to truly smart and decarbonized neighborhoods [65].

Together, these case studies illustrate the evolving trajectory from individual net-zero buildings to fully integrated community-scale systems that optimize renewable generation, storage, and distribution. They demonstrate that achieving true sustainability requires coordinated planning across buildings, infrastructure, and governance, transforming isolated efficiency measures into collective, low-carbon urban ecosystems.

7.5 Key Barriers to Achieving Net-Zero Energy Communities

Implementation of net-zero energy communities faces several key challenges. These challenges require systemic change, such as change of existing energy infrastructure, regulations, financial barriers, as well as societal challenges including resistance to change and equity issues [66].

The transition to net-zero communities needs to include a complete system, with a full understanding of its impacts [67]. The adoption of net zero is still quite slow in fast-growing markets as the idea is either new or in its early stages, making it still a reluctant topic to dive into [68, 69]. NZEC represents a very hopeful idea for the future and our environmental sustainability, however, the transition also has its challenges, one of them being the initial cost. Investment costs represent a barrier demotivating communities or investors into going forward into a whole new system [70]. Analysis made by the International Energy Agency and others display that Net Zero can cost up to $100 trillion globally [71, 72]. As mentioned within the UK policy and framework, the government has had an immense drive to push forward for net-zero emissions. However, their high ambitions face physical constraints, social and political resistance, as well as high costs [73–75] including the low-income families that may not be able to afford such luxuries [76].

7.5.1 Future Trend

Future trends in net-zero energy communities are related to adoption of emerging new approaches to enhance resilience and participation. These future pathways lead to the increase of local energy autonomy, improved energy efficiency, and strengthened community engagement that may include peer-to-peer energy trading, community microgrids, relevant regulations and standards, and social innovation.

Peer-to-Peer Energy Trading often involves individuals or communities with distributed energy sources, facilitating the exchange of clean energy, thus encouraging the production and consumption of renewable energy [77]. Researchers look to explore different techniques to price energy in community-based peer-to-peer (P2P) trading. One method includes using a supply–demand ratio for trading solar energy among neighbors with solar panels. Selling extra energy back into grids can encourage community members to use solar energy [78]. The traditional system of energy trading consists of a main plant sending electricity to a transmission grid, that is received by the consumer, it flows in one direction, and the customer purchases enough energy to satisfy the household needs [79]. Studies show that within the UK, a community with around 100 homes that use solar panels and private batteries, could cut energy bills by 30%, improving the community's self-sufficiency by 20% [80] and reducing their reliance on traditional non-renewable sources.

Community Microgrids introduce different kinds of renewable energy in different locations. This changes how people within the community use and think about energy, its success depends on the involvement and activeness of the community with tasks related to installing and operating the microgrid [81]. To achieve localized energy balance, microgrids have had a large demand toward energy systems which are distributed and decentralized [82]. Furthermore, community microgrids may be a great choice for achieving local energy balance and improving the integration of Renewable Energy Sources (RES) [83]. Since they are not dependent on a main power grid, these microgrid communities can continuously provide power which creates a consistent supply of energy for critical facilities and services [84, 85]. Community microgrids are beneficial in providing cost-saving benefits by minimizing the need for infrastructure upgrades, the localized energy and distribution results with more affordable efficient energy systems [86].

If community involvement is low, it could be detrimental to the functioning of the microgrid [87]. For successful community microgrids, their design needs to align with community behavior and social factors. Ultimately, the notion of community involvement links with the idea of net zero energy. Establishing microgrids can further progress more NZECs to appear and be adapted globally. The integration of microgrids into net-zero initiatives aligns with the overarching goals of achieving carbon balance.

Social innovation is developing new ideas and solutions which address social needs [88]. In energy, social innovation generally has a large acceptance amongst residents as it benefits local communities and is legitimized within the community and city organizations [89]. Through this awareness, social innovations can transform consumers into proactive citizens as they look for local solutions leading to easier energy transitions [90]. Social innovation can be applied to NZEC and its development for the future, as it allows constant engagement within communities and addresses social needs, thus promoting a just transition to a carbon balanced future. As already mentioned, the negative impacts of non-renewable energy sources, for example fossil fuels, made us question our actions and what we need to do to reduce GHG emissions. This leads us to developing and innovating RES to decrease pollution and create better net-zero communities for the future [91]. This process, however, is important regarding socio-economic challenges, through policy makers they need to be just and properly addressed to enable an achievable energy transition [92, 93]. It is said that to create a faster transition to sustainable energy systems, social innovation initiatives must be taken into consideration [94].

7.6 Conclusion

In conclusion, heading down the path of NZECs aims to create a balance with the energy we use through renewable sources and efficiency. Through case study of adapting solar-power energy seen in Pakistan, it displays how NZECs can provide sustainable electricity to places with no reach of power. However, challenges such as changing systems, costs, and community concerns need to be addressed. Future trends and ideas such as peer-to-peer energy trading and community microgrids offer endless possibilities and social innovations, whereas understanding people's needs and concerns is also crucial in establishing NZECs. Ultimately, achieving NZECs needs constant efforts, relevant standards and regulations adoption, adaptability to newer technology, and a commitment to address these environmental challenges.

References

1. Arif S, Taweekun J, Ali HM, Ahmed A, Bhutto AA (2023) Building resilient communities: techno-economic assessment of standalone off-grid PV powered net zero energy (NZE) villages. Heliyon 9. https://doi.org/10.1016/J.HELIYON.2023.E21426/ASSET/AD6489DA-BF2E-4B53-A578-543747F64539/MAIN.ASSETS/GR9.JPG

2. Carlisle N, Van Geet O, Pless S (2009) Definition of a zero net energy community. National Renewable Energy Laboratory
3. Ullah KR, Prodanovic V, Pignatta G, Deletic A, Santamouris M (2021) Technological advancements towards the net-zero energy communities: a review on 23 case studies around the globe. Sol Energy 224:1107–1126. https://doi.org/10.1016/J.SOLENER.2021.06.056
4. Raihan A (2023) The influences of renewable energy, globalization, technological innovations, and forests on emission reduction in Colombia. Innov Green Dev 2:100071. https://doi.org/10.1016/J.IGD.2023.100071
5. Selosse S (2019) Bioenergy with carbon capture and storage: how carbon storage and biomass resources potentials can impact the development of the BECCS. Bioenergy with carbon capture and storage: using natural resources for sustainable development, pp 237–256. https://doi.org/10.1016/B978-0-12-816229-3.00012-0
6. Raihan A, Pavel MI, Muhtasim DA, Farhana S, Faruk O, Paul A (2023) The role of renewable energy use, technological innovation, and forest cover toward green development: evidence from Indonesia. Innov Green Dev 2:100035. https://doi.org/10.1016/J.IGD.2023.100035
7. Tang D, Luo J, Sun H, Wu Z, Wang L (2023) A study on the impact of low carbon transition on green performance of industrial enterprises quasi-natural experiments based on low carbon city pilot. J Educ, Humanit Soc Sci 19:194–205. https://doi.org/10.54097/EHSS.V19I.11035
8. Fullerton D, Wolfram C (2012) The design and implementation of US climate policy. The design and implementation of U.S. climate policy, pp 197–209
9. Xu W, Zhang Y (2025) The impact of carbon taxes and carbon tax recovery on the Chinese economy: a green technological progress perspective. Sustainability 17:1700. https://doi.org/10.3390/SU17041700
10. Ma M (2024) The application of green GDP and its impact on global economy and environment: analysis of GGDP based on SEEA model
11. China's shift to consumption-led growth can aid green goals. https://www.imf.org/en/News/Articles/2022/01/27/cf-china-shift-to-consumption-led-growth-can-aid-green-goals. Accessed 15 Oct 2025
12. Cheng F, Luo H, Jenkins JD, Larson ED (2023) Inflation reduction act impacts on the economics of clean hydrogen and liquid fuels. Environ Sci Technol 57:15336–15347. https://doi.org/10.1021/acs.est.3c03063
13. Growth in CO_2 emissions leaves China likely to miss climate targets | Climate crisis | The Guardian. https://www.theguardian.com/environment/2024/feb/22/growth-in-co2-emissions-leaves-china-likely-to-miss-climate-targets. Accessed 15 Oct 2025
14. Skjærseth JB, Andresen S, Bang G, Heggelund GM (2021) The Paris agreement and key actors' domestic climate policy mixes: comparative patterns. Int Environ Agreements 21:59. https://doi.org/10.1007/S10784-021-09531-W
15. Joss S (2011) Eco-cities: the mainstreaming of urban sustainability—key characteristics and driving factors. Int J Sustain Dev Plan 6:268–285
16. Walker G (2011) The role for community in carbon governance. Wiley Interdisc Rev Clim Change 2:777–782
17. Energy-European commission. https://energy.ec.europa.eu/index_en. Accessed 30 Oct 2025
18. The European Green Deal-European commission. https://commission.europa.eu/strategy-and-policy/priorities-2019-2024/european-green-deal_en. Accessed 30 Oct 2025
19. Agyekum-Kwatiah K. Types of renewable energy sources By Ken. https://www.academia.edu/41523819/Types_of_renewable_energy_sources_By_Ken
20. Gergerli Z (2019) Solar energy, solar energy information. www.energymanagement.life
21. Pourasl HH, Barenji RV, Khojastehnezhad VM (2023) Solar energy status in the world: a comprehensive review. Energy Rep 10:3474–3493. https://doi.org/10.1016/J.EGYR.2023.10.022
22. Hachem-Vermette C, Guarino F, La Rocca V, Cellura M (2019) Towards achieving net-zero energy communities: investigation of design strategies and seasonal solar collection and storage net-zero. Sol Energy 192:169–185. https://doi.org/10.1016/J.SOLENER.2018.07.024

23. Lund H, Østergaard PA, Connolly D, Mathiesen BV (2017) Smart energy and smart energy systems. Energy 137:556–565. https://doi.org/10.1016/J.ENERGY.2017.05.123
24. WIND ENERGY | prince chandu—academia.edu. https://www.academia.edu/35210911/WIND_ENERGY. Accessed 06 Jan 2024
25. Comsan N. Wind energy for sustainable development. https://www.academia.edu/6333159/WIND_ENERGY_FOR_SUSTAINABLE_DEVELOPMENT
26. Ashiono A. Geothermal energy. https://www.academia.edu/37956029/Geothermal_Energy
27. Holland J. Electricity, the benefits of geothermal energy. https://www.academia.edu/10618732/Electricity_the_Benefits_of_Geothermal_Energy
28. Avtar R, Kumar P (2016) Geothermal energy as a major source of renewable energy—learning from Asian neighbours. Adv Geoth Energy
29. Demirbas A (2008) Importance of biomass energy sources for Turkey. Energy Policy 36:834–842
30. McKendry P (2002) Energy production from biomass (part 1): overview of biomass. Bioresour Technol 83:37–46. https://doi.org/10.1016/S0960-8524(01)00118-3
31. Renewables Global Status Report—REN21. https://www.ren21.net/reports/global-status-report/. Accessed 30 Oct 2025
32. Home-Mission Innovation. https://www.mission-innovation.net/. Accessed 30 Oct 2025
33. Renné DS (2022) Progress, opportunities and challenges of achieving net-zero emissions and 100% renewables. Solar Compass 1:100007. https://doi.org/10.1016/J.SOLCOM.2022.100007
34. Nouban F, Abazid M (2017) Plastic degrading fungi Trichoderma viride and Aspergillus nomius isolated from local landfill soil in Medan, Iopscience.Iop.Org 8:68–74. https://doi.org/10.1088/1755-1315/140/1/012106/PDF
35. Technology and innovation pathways for zero-carbon-ready buildings by 2030—Analysis—IEA. https://www.iea.org/reports/technology-and-innovation-pathways-for-zero-carbon-ready-buildings-by-2030. Accessed 30 Oct 2025
36. Energy Efficiency Policy Toolkit 2025—analysis—IEA. https://www.iea.org/reports/energy-efficiency-policy-toolkit-2025. Accessed 30 Oct 2025
37. Nearly-zero energy and zero-emission buildings—European commission. https://energy.ec.europa.eu/topics/energy-efficiency/energy-performance-buildings/nearly-zero-energy-and-zero-emission-buildings_en. Accessed 30 Oct 2025
38. Government of the UK: net zero strategy: build back greener. Gov.Uk. 368 (2021)
39. Walker A, Carver D. Government policy on reaching Net Zero by 2050. https://commonslibrary.parliament.uk/research-briefings/cdp-2023-0124/
40. The ten point plan for a green industrial revolution—GOV.UK. https://www.gov.uk/government/publications/the-ten-point-plan-for-a-green-industrial-revolution. Accessed 07 Jan 2024
41. Zehri C, El Amin BM, Kadja A, Inaam Z, Sekrafi H (2024) Decent work, financial inclusion, and economic growth: analysis under the SDG 8. Heritage Sustain Dev. 6:197–218. https://doi.org/10.37868/HSD.V6I1.370
42. 2022 progress report to parliament—climate change committee. https://www.theccc.org.uk/publication/2022-progress-report-to-parliament/. Accessed 07 Jan 2024
43. Copeland C, MacKerron G, Foxon TJ (2023) Futures for Findhorn: exploring challenges for achieving net zero in an ecological intentional community. Futures 149:103155. https://doi.org/10.1016/J.FUTURES.2023.103155
44. Tarazona-Romero BE, Socha-Rojas ND, Ascanio-Villabona J, Kafarov V, Castillo-Leon NY (2024) Performance evaluation of solar thermal collectors in Colombian thermal floors by dynamic simulation. Sustain Eng Innov 6:421–432. https://doi.org/10.37868/SEI.V6I2.ID394
45. Bianchi A, Ginelli E (2018) The social dimension in energy landscapes. City Territory Archit 5
46. Alkemade F, de Bruin B, El-Feiaz A, Pasimeni F, Niamir L, Wade R (2024) Social tipping dynamics in the energy system. Earth Syst Dyn 15:485–500. https://doi.org/10.5194/esd-15-485-2024
47. HM Government: powering up Britain. UK Goverment (2023)

48. Powering up Britain: energy security plan—GOV.UK. https://www.gov.uk/government/publications/powering-up-britain/powering-up-britain-energy-security-plan. Accessed 07 Jan 2024
49. Clean air zone framework—GOV.UK. https://www.gov.uk/government/publications/air-quality-clean-air-zone-framework-for-england/clean-air-zone-framework. Accessed 07 Jan 2024
50. Clean air zones—GOV.UK. https://www.gov.uk/guidance/driving-in-a-clean-air-zone. Accessed 07 Jan 2024
51. Monitor frameworks for climate and energy targets | New Climate Institute. https://newclimate.org/resources/publications/monitor-frameworks-for-climate-and-energy-targets. Accessed 30 Oct 2025
52. Enabling framework for renewables. https://energy.ec.europa.eu/topics/renewable-energy/enabling-framework-renewables_en. Accessed 30 Oct 2025
53. Rodrigues L, Gillott M, Waldron J, Cameron L, Tubelo R, Shipman R, Ebbs N, Bradshaw-Smith C (2020) User engagement in community energy schemes: a case study at the Trent Basin in Nottingham, UK. Sustain Cities Soc 61:102187. https://doi.org/10.1016/J.SCS.2020.102187
54. D'Oca S, Hong T, Langevin J (2018) The human dimensions of energy use in buildings: a review. Renew Sustain Energy Rev 81:731–742. https://doi.org/10.1016/J.RSER.2017.08.019
55. Warner KJ, Jones GA (2017) A population-induced renewable energy timeline in nine world regions. Energy Policy 101:65–76. https://doi.org/10.1016/J.ENPOL.2016.11.031
56. Energy Policy (2017) 101:65–76. https://scholar.google.com/scholar?q=Energy%20Policy%202017%2C%20101%2C%2065-76. Accessed 07 Jan 2024
57. Mirjat NH, Uqaili MA, Harijan K, Valasai GD, Shaikh F, Waris M (2017) A review of energy and power planning and policies of Pakistan. Renew Sustain Energy Rev. 79:110–127. https://doi.org/10.1016/J.RSER.2017.05.040
58. Arif S, Taweekun J, Ali HM, Ahmed A, Bhutto AA (2023) Building resilient communities: techno-economic assessment of standalone off-grid PV powered net zero energy (NZE) villages. Heliyon 9:e21426. https://doi.org/10.1016/J.HELIYON.2023.E21426
59. Bhutto AW, Bazmi AA, Zahedi G (2011) Greener energy: issues and challenges for Pakistan—biomass energy prospective. Renew Sustain Energy Rev 15:3207–3219. https://doi.org/10.1016/J.RSER.2011.04.015
60. Fuso Nerini F, Howells M, Bazilian M, Gomez MF (2014) Rural electrification options in the Brazilian Amazon: a multi-criteria analysis. Energy Sustain Dev 20:36–48. https://doi.org/10.1016/J.ESD.2014.02.005
61. Givler T, Lilienthal P (2005) Using HOMER software, NREL's micropower optimization model, to explore the role of gen-sets in small solar power systems; case study: Sri Lanka. https://doi.org/10.2172/15016073
62. Yang S-H, Boo C-J, Kim H-C (2012) Optimization of stand-alone hybrid power systems using HOMER program. J Korean Solar Energy Soc 32:11–18. https://doi.org/10.7836/KSES.2012.32.2.011
63. Sougkakis V, Lymperopoulos K, Nikolopoulos N, Margaritis N, Giourka P, Angelakoglou K (2020) An investigation on the feasibility of near-zero and positive energy communities in the Greek context. Smart Cities 3:362–384
64. Belmar F, Baptista P, Neves D (2023) Modelling renewable energy communities: assessing the impact of different configurations, technologies and types of participants. Energy Sustain Soc 18
65. Moghaddasi H, Culp C, Vanegas J (2021) Net zero energy communities: integrated power system, building and transport sectors. Energies (Basel) 14
66. Heti K, Silvanskaya G, Khambir V, Harkava V, Ivanchenko K (2025) The analysis of engineering adaptation to global challenges and strategic planning for the future. Sustain Eng Innov 7:137–146. https://doi.org/10.37868/SEI.V7I1.ID453
67. Hsu A, Logan K, Qadir M, Booysen MJ (Thinus), Montero AM, Tong K (KK), Broadbent G, Wiedmann T, Sin Woon VK, Good C, Colding J, Foliente G, Kılkış Ş (2022) Opportunities and barriers to net-zero cities. One Earth 5:739–744. https://doi.org/10.1016/J.ONEEAR.2022.06.013

68. Martiskainen M, Kivimaa P (2019) Role of knowledge and policies as drivers for low-energy housing: case studies from the United Kingdom. J Clean Prod 215:1402–1414. https://doi.org/10.1016/J.JCLEPRO.2019.01.104
69. Persson J, Grönkvist S (2015) Drivers for and barriers to low-energy buildings in Sweden. J Clean Prod 109:296–304. https://doi.org/10.1016/J.JCLEPRO.2014.09.094
70. BloombergNEF's: energy transition investment trends 2022. Technical report (2022)
71. IEA (2021) Net zero by 2050: a roadmap for the global energy sector—international energy agency. International Energy Agency
72. Managerial and financial barriers to the net-zero transition | Mirabelle Muûls—Academia.edu. https://www.academia.edu/93523865/Managerial_and_Financial_Barriers_to_the_Net_Zero_Transition. Accessed 07 Jan 2024
73. Wilson C, Tyfield D (2018) Critical perspectives on disruptive innovation and energy transformation. Energy Res Soc Sci 37:211–215. https://doi.org/10.1016/J.ERSS.2017.10.032
74. Nelson S, Allwood JM (2021) Technology or behaviour? Balanced disruption in the race to net zero emissions. Energy Res Soc Sci 78:102124. https://doi.org/10.1016/J.ERSS.2021.102124
75. McDowall W (2018) Disruptive innovation and energy transitions: is Christensen's theory helpful? Energy Res Soc Sci 37:243–246. https://doi.org/10.1016/J.ERSS.2017.10.049
76. Smith L, Tyers R. Clean air zones, low emission zones and the London ULEZ. https://commonslibrary.parliament.uk/research-briefings/cbp-9816/
77. Denysiuk R, Lilliu F, Recupero D, Vinyals M (2020) Peer-to-peer energy trading for smart energy communities. Proceedings of the 12th international conference on agents and artificial intelligence
78. Tushar W, Saha TK, Yuen C, Morstyn T, McCulloch MD, Poor HV, Wood KL (2019) A motivational game-theoretic approach for peer-to-peer energy trading in the smart grid. Appl Energy 243:10–20. https://doi.org/10.1016/J.APENERGY.2019.03.111
79. Feasibility Study-WREN. https://www.wren.uk.com/energy-equality/feasibility-study. Accessed 17 Jan 2024
80. Qin H, Pan W (2020) Energy use of subtropical high-rise public residential buildings and impacts of energy saving measures. J Clean Prod 254:120041. https://doi.org/10.1016/J.JCLEPRO.2020.120041
81. Eklund M, Khalilpour K, Voinov A, Hossain MJ (2023) Understanding the community in community microgrids: a conceptual framework for better decision-making. Energy Res Soc Sci 104:103260. https://doi.org/10.1016/J.ERSS.2023.103260
82. Hirsch A, Parag Y, Guerrero J (2018) Microgrids: a review of technologies, key drivers, and outstanding issues. Renew Sustain Energy Rev 90:402–411. https://doi.org/10.1016/J.RSER.2018.03.040
83. Ton DT, Smith MA (2012) The U.S. department of energy's microgrid initiative. The Electricity J 25:84–94. https://doi.org/10.1016/J.TEJ.2012.09.013
84. Hicks J, Ison N (2018) An exploration of the boundaries of community in community renewable energy projects: navigating between motivations and context. Energy Policy 113:523–534. https://doi.org/10.1016/J.ENPOL.2017.10.031
85. Warneryd M, Håkansson M, Karltorp K (2020) Unpacking the complexity of community microgrids: a review of institutions roles for development of microgrids. Renew Sustain Energy Rev 121:109690. https://doi.org/10.1016/J.RSER.2019.109690
86. Ruggiero S, Onkila T, Kuittinen V (2014) Realizing the social acceptance of community renewable energy: a process-outcome analysis of stakeholder influence. Energy Res Soc Sci 4:53–63. https://doi.org/10.1016/J.ERSS.2014.09.001
87. Cuesta MA, Castillo-Calzadilla T, Borges CE (2020) A critical analysis on hybrid renewable energy modeling tools: an emerging opportunity to include social indicators to optimise systems in small communities. Renew Sustain Energy Rev 122:109691. https://doi.org/10.1016/J.RSER.2019.109691
88. Spotlight series: cities as ecosystems for social innovation—NetZeroCities. https://netzerocities.eu/2023/03/17/spotlight-series-cities-as-ecosystems-for-social-innovation/. Accessed 07 Jan 2024

89. Sovacool BK, Brugger H, Brunzema I, Dańkowska A, Wemyss D, Vernay AL, Betz R, Avelino F, de Geus T, Dembek A, Duetschke E, Hielscher S, Iskandarova M, Müller L, Musiolik J, Ranville A, Schleich J, Stasik A, Strumińska-Kutra M, Winzer C, Wittmayer J, Rogge KS (2023) Social innovation supports inclusive and accelerated energy transitions with appropriate governance. Commun Earth Environ 4. https://doi.org/10.1038/S43247-023-00952-W
90. Hewitt RJ, Bradley N, Compagnucci AB, Barlagne C, Ceglarz A, Cremades R, McKeen M, Otto IM, Slee B (2019) Social innovation in community energy in Europe: a review of the evidence. Front Energy Res 7. https://doi.org/10.3389/FENRG.2019.00031
91. Selvakkumaran S, Ahlgren EO (2020) Impacts of social innovation on local energy transitions: diffusion of solar PV and alternative fuel vehicles in Sweden. Glob Transit 2:98–115. https://doi.org/10.1016/J.GLT.2020.06.004
92. Hölsgens R, Lübke S, Hasselkuß M (2018) Social innovations in the German energy transition: an attempt to use the heuristics of the multi-level perspective of transitions to analyze the diffusion process of social innovations. Energy Sustain Soc 8. https://doi.org/10.1186/S13705-018-0150-7
93. Lennon B, Dunphy NP, Sanvicente E (2019) Community acceptability and the energy transition: a citizens perspective. Energy Sustain Soc 9. https://doi.org/10.1186/S13705-019-0218-Z/.
94. Hewitt RJ, Bradley N, Compagnucci AB, Barlagne C, Ceglarz A, Cremades R, McKeen M, Otto IM, Slee B (2019) Social innovation in community energy in Europe: a review of the evidence. Front Energy Res 7. https://doi.org/10.3389/FENRG.2019.00031/FULL
95. Abed FT, Abed NK, ALRubeei IRN, Alrikabi ARH (2024) Using genetic algorithm for optimal sizing of stand-alone hybrid energy system. Heritage Sustain Dev 6:287–300. https://doi.org/10.37868/HSD.V6I1.501

Chapter 8
Shaping the Future of Zero Energy Buildings

8.1 Introduction

The global population growth and rapid development of the economy lead to greater demand for energy. Currently, majority of energy comes from fossil fuels, which contributes to global warming and climate change, increasing sea and land temperatures, and causing more frequent heat waves. According to several research papers, the global temperature is expected to increase by 1.5 °C by the year 2030 [1]. Therefore, global warming has become one of the main concerns of global society in recent years.

Buildings use energy from fossil fuels to provide occupants with comfort. As they represent one of the most significant contributors for energy use, this also ensures a great opportunity for reducing energy consumption and switching to self-sustainable forms to significantly reduce greenhouse gas emissions [2]. Along with harming the environment, as proven by many studies, exposure to increasing temperatures is extremely harmful to human health. High temperatures increase the risk of various diseases [3], and allergies. To achieve global climate change goals and preserve the earth and human health, it is necessary to reduce greenhouse gas emissions as soon as possible [4].

Driven by these concerns, many countries are taking various measures to raise awareness of global warming and its consequences. To reduce the impact of global climate change, countries are adopting new laws and promoting awareness of global warming by encouraging the adoption of Zero Energy Buildings (ZEBs) and the renovation of existing structures. ZEBs produce energy in the same amount as they consume, therefore represent a good option for changing to a better future. Despite the urgent need for energy-efficient renovations, the renovation rate of energy-inefficient buildings is still insufficient [5].

This chapter explores the current trends, technological innovations, policy frameworks, and future directions that will shape the evolution of ZEBs in the coming decades.

B. Duraković, *Advancing Zero Energy Buildings*, Green Energy and Technology,
https://doi.org/10.1007/978-3-032-16601-2_8

8.2 Technologies Driving ZEBs

The first consideration in designing energy-efficient buildings should be a well-structured passive design with an emphasis on excellent thermal and living comfort. The focus should be on climate, orientation, geometry and thermal characteristics of the building. During the entire life cycle of the building, it is important to take into account the protection of the environment, achieved by using recycled materials and as little energy as possible during production [6].

Energy efficiency is a primary goal during the construction of zero-energy buildings, precisely so that the remaining energy demand can be met from renewable sources [7]. In other words, the annual net balance between the total energy consumed by buildings and produced energy should be zero. During the design phase, it is necessary to introduce several energy efficiency measures for the building to become nearly ZEB, but also to produce as much energy as possible from renewable sources [8]. This is achievable adopting passive and active strategies and construction methods for renewable energy sources from planning to completion of the building. Passive design strategies cover site planning, building orientation, daylight control, materials, facade design and construction techniques. Active design strategies include the optimization of HVAC systems, energy-efficient appliances and equipment. Figure 8.1 represents some of the common ZEB energy efficient technologies applicable for passive and active design.

8.2.1 *Building Envelope and Systems Innovations*

Passive technologies play a fundamental role in enhancing the energy efficiency of building envelopes by minimizing the need for active mechanical systems. The

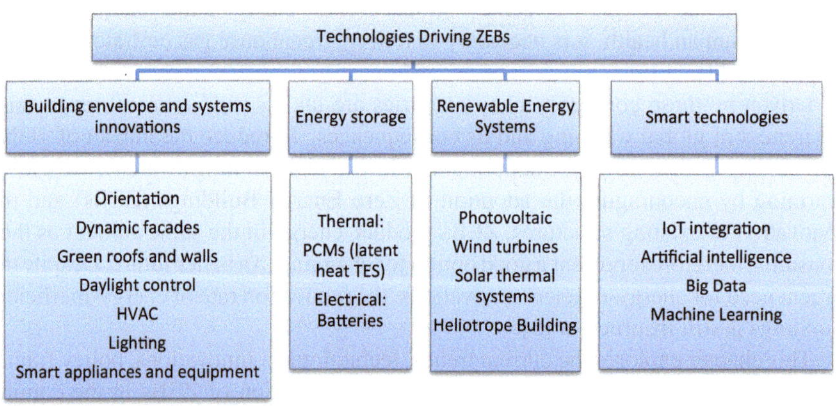

Fig. 8.1 ZEB energy efficient technologies

passive design systems harvest natural energy flows including solar radiation, wind, and thermal mass to obtain occupant thermal comfort by regulating indoor temperatures. Key technologies that enable passive design are built in optimized *building orientation* that serves to maximize solar gain in winter and minimize it in summer, the use of *high-performance insulation materials* to reduce heat transfer, and strategic placement of *windows* and *shading devices* to control daylight and prevent overheating [9]. Furthermore, natural ventilation systems allow airflow without installment of mechanical devices, while *thermal mass materials absorb*, store, and release heat when needed to stabilize indoor temperatures [10]. Additional insolation is achieved by integrating green roofs and walls into building design to mitigate heat island effects [11]. The integration of passive design systems allows buildings to significantly reduce energy consumption for heating, cooling, lighting, and lower operational costs and carbon emissions while creating healthier indoor environments [9]. Automated control technologies, network sensors and instrumentation, energy management and information systems are just a few of the energy efficiency technologies [9]. A good few of the new technologies that help us achieve these goals are dynamic facades, green roofs and walls and daylight control.

Building orientation is about how the entire building is positioned on a site relative to the sun's path (north, south, east, west). It is a macro-level decision that affects overall solar heat gain, natural heating, and cooling potential throughout the year. It is observed that optimal orientation can reduce annual energy use up to ~ 49% compared to less optimal orientations [12]. *Daylight Control* is about how natural light enters and how it is managed inside the building through design features like window placement, shading devices (overhangs, blinds), glazing types, and interior layouts to maximize useful daylight, reduce glare, and avoid overheating. It is reported in the literature that optimal window orientation and shading devices could improve daylight by up to 70%, but at the same time can reduce energy demand by nearly 59% [13].

The *dynamic facade* (DF) technologies are a huge opportunity for the emission of greenhouse gases and the reduction of the energy demand of buildings while simultaneously improving the quality of the surroundings [14]. Dynamic facades make a significant contribution to energy efficiency and livability which is largely due to their interaction with external factors. Dynamic Facades combined with adaptive systems are a key part of a study [15] that is investigating climate zones and their impact on the performance of dynamic facades. It is concluded that moderate climates with dynamic facades, in terms of energy efficiency, benefit the most from daylight, while dry climates benefit more from thermal comfort [15].

DFs aim to provide residents with better indoor circumstances, especially in terms of thermal comfort [16], thus lowering building total energy use [17]. A variety of terminologies and classification systems have been used for dynamic facades [18], based on goals and different elements such as technology, response time, spatial scale, visibility, degree of adaptation and work style [19]. The results of these factors are crucial for the optimal design and execution of dynamic facades. According to the research [15], DFs can be classified in different types including: Kinetic façade, Shading system, Ventilated façade, Double-skin façade, Thermal energy storage façade, Dynamic coating system, Radiant wall façade, Thermoelectric façade, and Pneumatic façade.

These diverse types of dynamic façades demonstrate the wide range of innovative approaches available to enhance building performance, energy efficiency, and occupant comfort through adaptable and responsive envelope technologies. Selecting the appropriate façade type depends on specific climate conditions, building use, and design goals.

Green roofs, also known as vegetative or living roofs, are an innovative passive design strategy that involves growing vegetation on building rooftops to improve thermal insulation, manage stormwater, and enhance urban biodiversity [20]. A green roof generally consists of up to four main layers including waterproof membranes, drainage films, soil, and vegetation. The roof plays one of the most influential roles in the thermal properties of buildings, namely the roof absorbs solar radiation which significantly contributes to the increase in heat inside the buildings. During a sunny, cloudless day, the roof can receive up to 1 kW/m^2 of solar radiation, of which 20% to 95% is typically absorbed [21]. The soaked up heat is shifted into the building, which leads to greater internal temperatures and discomfort for the residents, and an increase in their need for air conditioning, therefore a greater need for energy consumption [22].

In addition, green roofs can be placed in two categories, intensive green roofs and extensive green roofs [21]. Intensive green roofs are designed for aesthetics and entertainment, requiring extra structure to support larger plant life, high maintenance and investments, also a deep soil. Extensive green roofs are cost-effective, requiring less maintenance, with thin and shallow soil layer, and serving a more functional purpose compared to intensive green roofs [21].

Advanced *heating, ventilation, and air conditioning* (HVAC) technologies, such as high-efficiency heat pumps, energy recovery ventilation, and smart controls, play a pivotal role in achieving ZEB standards, as they significantly influence energy consumption and indoor environmental quality. For example, integrating energy recovery ventilation systems can reduce energy use by up to 17.4% in HVAC and overall building energy consumption by 9.7% in residential settings [23]. The ground-source heat pumps (GSHPs) can achieve energy savings from 24.3% to 39.2% in heating-dominated climates. Additionally, the incorporation of artificial intelligence (AI) algorithms into HVAC systems can reduce energy demand by up to 40%. Overall, these advancements in HVAC technologies are crucial in contributing to the realization of ZEBs [24].

Lighting systems are another technology driving ZEBs through the integration of energy-efficient technologies and intelligent control mechanisms. Light Emitting Diodes (LEDs) is a major element in this transition due to its high luminous efficiency and longevity [25]. If LEDs combined with daylight harvesting techniques energy consumption can be reduced by up to 69% [26]. If they are used with smart lighting technologies, energy savings can be achieved up to 80% under certain occupancy scenarios [27]. LEDs have an energy loss in the form of heat of 70%, but it is still an advantage over other forms of light bulbs such as incandescent bulbs that emit heat and infrared radiation, making them lose 97% of their energy [28, 29].

Another energy efficient lighting system is composed of compact fluorescent lamps (CFL) [30]. Controls such as dimmers, occupancy sensors, timers, and other advanced lighting controls allow us to alter lighting levels based on time of day or natural light levels, ensuring that lights are used only when and where they are needed, maximizing energy savings and optimizing lighting efficiency [31]. Advanced controls can reduce energy consumption and lower utility costs just by controlling light levels.

Smart appliances and equipment can optimize energy consumption through smart automation and real-time data analytics. An example of these technologies includes smart thermostats, smart washing machines and dryers, smart refrigerators, smart HVAC systems, smart lighting systems etc. These systems use sensors and artificial intelligence (AI) to adjust ZEBs operations based on energy availability, occupancy, and environmental conditions. Based on the user preference patterns, smart thermostats can learn to optimize energy use and achieve the energy reduction by up to 10% [32], while the AI-powered occupancy detection systems can achieve the energy reduction between 11.1 and 13.1% [33]. The integration of renewable energy sources with smart appliances enhances energy efficiency of ZEBs. If the smart appliances are integrated with Building Energy Management Systems (BEMS) it can facilitate real-time monitoring and control [34, 35].

The latest developments in electrical power delivery are smart grid systems. The combination of advanced information technologies and traditional electricity transmission systems enable better control and efficiency of electricity [36]. By optimizing the energy efficiency of smart networks with the help of smart meters and devices, better energy efficiency can be achieved. Smart meters are high precision instruments that monitor electricity consumption in real time and provide better efficiency of appliances linked to the smart grid [37]. Load management program implementation is another technique for energy efficiency in a smart grid. These programs manage peak load periods and allow utilities to adjust energy prices to encourage off-peak periods [38] by shifting energy demand and minimizing cost. Energy storage and load reduction also help in reducing the costs as well as the use of additional (sustainable) energy sources [39].

8.2.2 Energy Storage Systems

Energy storage system (ESS) is considered as one of the key technologies in optimizing building energy performance and reducing reliance on the grid. The storage systems use the excess energy generated during periods of low demand and released it during peak consumption hours. The storage systems enable efficient storage and utilization of energy either as thermal or electrical. Therefore, ESS supports the integration of renewable energy sources and contributes to the overall resilience of the building energy infrastructure.

The energy is stored in various ways such as thermal, grid flow batteries or hydrogen storage [40, 41] as one of the most important and effective methods for successfully integrating renewable sources of electricity into the grid. Renewable energy storage helps overcome unpredictable challenges, and it does this by efficiently storing excess energy during times of abundant energy, in the case of solar power plants, sunny days, and in the case of wind power plants, of course, windy days. And during times of big appetite or poor production, it can release energy [42]. As a result, these technologies provide a more reliable and stable energy supply, improve grid stability and enable better utilization of renewables.

8.2.3 Renewable Energy Systems Integration

Renewable energy systems is another key technology in achieving ZEB's core objective. However, the integration of these technologies, such as solar photovoltaics, wind turbines, geothermal systems, and heliotropes within the ZEBs operational framework requires careful design, smart energy management, and advanced storage solutions [43].

Building-integrated photovoltaic (BIPV) systems have recently garnered significant attention as an effective approach for sustainable energy provision [44]. Photovoltaic (PV) is a system of capturing sunlight, converting it into direct current, which is what powers the building [44]. Photovoltaic cells are taking the place of conventional building materials for roof panels, walls and window glazing [45]. Integrating photovoltaic systems into buildings is to provide energy for different building applications such as Lighting, Heating, Ventilation, and Air Condition which makes new buildings energy efficient.

Deep learning, as a subset of artificial intelligence, has proven to be an effective method for enlarging the efficiency of interconnected systems. PV systems can use deep learning algorithms to maximize energy harvest under all conditions to predict solar radiation and production patterns, as demonstrated in a study [46] that found that improving PV forecasting can significantly increase the efficiency of solar power generation. This depends on geographical (location) and temporal (weather) influences, precisely because of their unpredictability. A new forecasting model is presented that shows that solar systems can achieve improved accuracy and efficiency by correctly grouping data. Therefore, PV systems are optimistic technologies to meet requirements of producing the energy to support ZEBs [47, 48].

Building-integrated wind turbines (BIWT) are divided into groups according to the number of turbines, their location, and their purpose. The first category includes one or two large wind turbines on top of tall buildings. The second group is placement of turbines between the passage of two buildings, and the third group is placement of the turbines in specially constructed holes inside a building. In the second category, multiple tiny wind turbines may be on the corners or ridge of a construction. Small wind turbines are generally less cost-effective than large-scale ones. This type of building-integrated wind turbine is considered a suitable and economical method [49].

Solar electric windows combine photovoltaic technology with smart window features to generate electricity while regulating sunlight, which results in great light transmission, efficient energy conversion and thermal comfort. They reduce energy consumption and carbon footprint giving the contribution to global climate goals while improving energy efficiency in buildings [50]. *Solar thermal system* is a mature and cost-effective solution to convert sunlight into usable heat for domestic hot water, space heating, or combined uses. Globally, approximately 86% of solar thermal deployments are dedicated to domestic hot water, an provide about 8–13% of a building domestic hot water energy need [51]. Emerging hybrid photovoltaic-thermal (PVT) technologies further enhance performance of solar thermal systems, which at the same time generate electricity and heat by cooling the PV modules [51].

Heliotrope Building is an architectural masterpiece made in Germany in 1994 as one of the first "Plus Energy Buildings" to protect the environment and strengthen the economy. The energy collected by collectors built into the house is used for electricity, heating and hot water. The house rotates on its axis to follow the Sun and collect as much energy as possible. Depending on the time of year, the Heliotrope produces four to six times more energy than it uses [52].

8.2.4 Smart Technologies

Artificial intelligence is a hot topic today, with numerous publications and discussions on its definition and application. It is evident that people interpret AI differently, such as some life forms that can surpass human intelligence, while others consider AI to be any kind of data processing technology. AI is designed to help people and is already doing so in various ways, such as providing cleaner air, optimizing energy systems and reducing energy consumption [53]. One of the most important subjects when it comes to AI and energy efficiency is Big Data and Machine Learning.

Big Data is a technology that helps in collecting and processing a huge volume of structured and unstructured data in real time [54, 55]. Energy use is massive, especially in buildings, where everyone needs large quantities at affordable prices. This task was previously quite difficult to perform, but the creation of tools to process these big data sets has made the process a much easier. Accurate prediction of energy consumption is crucial in the development of energy-efficient buildings, therefore data-driven methods provide more accurate prediction and have therefore become popular for estimating energy consumption in buildings [56]. The usefulness of Big Data can also be manifested through smart meters, which are being installed in electricity networks. Classic meters collect consumption data once a month, while smart meters do so every 15 min [57]. In this way, such meters could provide very useful information and help in monitoring the quality of electricity, and at the same time in working decisions to reduce interruptions and losses [58]. One of the examples that prove usefulness of Big Data in ZEB has taken place in China and it investigates how solar panels for building facades generate energy, and how sunlight affects their performance. It also explores the interaction between solar energy and building energy consumption. Findings are applied to the design. It uses Big Data program to convert the original building into an energy efficient structure [59].

Machine learning is a subset of artificial intelligence (AI), and just as a person becomes more intelligent by learning more, the same principle applies to AI systems. The system can learn from data and make decisions without human intervention. In simple terms, machine learning teaches AI how to learn [60]. For machine learning to be effective, a lot of different types of information and different sources need to be available, which can relate to Big Dana concept. There are a lot of different ways machine learning is used to improve energy efficiency of zero energy building, but one of the best is by its optimization by heat recovery in a co-generation system, which produces electricity, heating and cooling [61]. The main goal of the study was to minimize the costs of living and maximize energy efficiency. Four different Australian towns were analyzed and studied according to their annual ambient temperatures. The results showed that the system meets the requirements for electricity for 294 apartments and heating for as many as 208 apartments [61].

The use of artificial intelligence in the energy industry has rapidly become globally widespread, mostly thanks to its ability to use predictive analytics algorithms that estimate the production of renewable energy with astonishing accuracy [62]. This predictive analytics also acts in the optimization, distribution and consumption of energy, and directing operations in the energy value chain [63]. The urban AI industry needs to refocus on energy efficiency issues from an economic and structural upgrade perspective [64]. Namely, there has been a shift from the old dilemma, moving away from the traditional debate between pollution and growth towards high-tech iteration and inclusively aligned social sustainability [65]. Although AI has many advantages in the preservation and advancement of sustainable technologies, it can also be one of the harmful factors and lead to significant consumption of primary energy and concerns for the environment [66]. The positive impacts of AI on energy efficiency focusing on the city scale have been examined by a few studies [67]. The potential of artificial intelligence in solving the above mentioned problem in the context of the rapid development of green cities is and it stays condescending [65].

8.2.5 Summary of Benefits and Challenges

Despite potential of technologies driving ZEBs, they often face challenges such as high initial costs, maintenance requirements, and different efficiency under various conditions. Understanding the benefits and limitations of these technologies, it is very important to select the most effective solutions in sustainable building design. Table 8.1 outlines the key technologies driving the ZEB development, highlighting their benefits and challenges.

Referring to Table 8.1, it can be observed that the main technologies supporting the ZEBs development include dynamic facades and green roofs to smart HVAC systems, LED lighting, smart grids, and renewable energy storage solutions. Overall, each of these technologies contributes uniquely to enhancing building energy performance.

Table 8.1 Benefits and challenges of technologies driving ZEBs

Aspect	Description	Benefits	Challenges
Dynamic facades	The concept of dynamic facade technologies is a huge opportunity for the emission of greenhouse gases and the reduction of the energy demand of buildings while simultaneously improving the quality of the surroundings [14]	DFs aim to provide residents with better indoor circumstances, especially in terms of thermal comfort [16], thus lowering building total energy use [17]	Costs, maintenance
Green roof	A green roof generally consists of up to 4 main layers; waterproof membranes, drainage films, soil, and vegetation [21]	Some of the benefits of green roofs include; "reducing energy consumption, mitigating the urban heat island effect, sequestering carbon dioxide, reducing stormwater runoff and flood risk, and improving air quality and urban living conditions" [21]	High maintenance and investments
HVAC	HVAC systems can significantly reduce energy consumption in buildings, especially HVAC systems that focus and operate based on different thermal zones. They are also helped by smart meter systems that enable the categorization of rooms according to the kind of user and their energy consumption habits [68]	The study [68], showed that constant energy users have the lowest efficiency during the heating season, and that night-time users show low energy efficiency when the dry bulb temperature is above 0 °C	Biggest energy consumer
LED lighting	LED lighting has expanded globally in the twenty-first century, thanks primarily to its energy efficiency but also to its long lifespan [25]	Although it has many virtues such as low voltage, reliability and high energy efficiency, it has an energy loss in the form of heat of 70%, but it is still an advantage over other forms of light bulbs such as incandescent bulbs that emit heat and infrared radiation, making the lose 97% of their energy [28, 29]	Higher costs

(continued)

Table 8.1 (continued)

Aspect	Description	Benefits	Challenges
Smart grid	Latest developments in electrical power delivery are smart grid systems. The combined advances of information technology and traditional electricity transmission systems allow for better control and efficiency of electricity [36]	By optimizing the energy efficiency of smart networks with the help of smart meters and devices we can achieve better energy efficiency. Smart meters serve for users to closely monitor electricity consumption in real time, we classify them as high precision instruments. Appliances are linked to the smart grid which gives us better efficiency [37]	Higher costs
Renewable energy storage	Efforts are being made to store energy in various ways such as grid flow batteries and hydrogen storage [40, 41]	Renewable energy storage helps overcome unpredictable challenges, and it does this by efficiently storing excess energy during times of abundant energy, in the case of solar power plants, sunny days, and in the case of wind power plants, of course, windy days. And during times of big appetite or poor production, it has the ability to release energy [42]	Higher costs

Despite these benefits, these technologies face certain challenges such as high costs, high maintenance demands, and technical limitations.

8.3 Circular Economy and Sustainable Materials

Incorporating recycled and locally sourced materials into ZEB construction reduces embodied carbon, aligning with circular economy principles. The concept of the circular economy is a regenerative system that emphasizes reuse, recycling, and waste minimization. The traditional economy is not a regenerative system, it is a linear system that extract natural raw materials, makes products and dispose them. This means that materials should be carefully selected in design phase so that, at the end of their life, components can be easily disassembled and repurposed or recycled rather than sent to landfills. Therefore, the circular economy model offers promising pathways toward sustainable construction practices that benefit the environment and society [69]. Table 8.2 shows key characteristics of the traditional linear economy with circular economy in construction.

Table 8.2 Linear vs. circular economy in construction comparisons

Aspect	Linear economy	Circular economy
Resource use	Extracts virgin materials	Utilizes recycled, renewable, and reclaimed inputs
Design approach	Designed for single use	Designed for disassembly and reuse
Waste management	High levels of construction and demolition waste	Waste minimized and materials reintegrated
End of life	Demolition and disposal in landfills	Materials recovered, reused, or recycled
Environmental impact	High carbon footprint	Reduced emissions and environmental impact
Economic model	Linear: take-make-dispose	Regenerative and restorative

Table 8.2 demonstrates various aspects of the traditional linear economy and circular economy models. Circular economy is built on resource efficiency that implies reuse and recycling. In context of buildings, this model enhances resilience and sustainability and encourage building design for disassembly and material recovery [70]. Below presented Table 8.3 summarizes key principles associated with circular economy in the construction sector.

Referring to Table 8.3 it can be observed that the adoption of circular economy principles brings many environmental, economic, and social benefits. Environmental benefits are reflected in reducing the raw materials extraction, minimizing the construction and demolition waste [76]. Economic benefits include cost savings, lower dependency on volatile raw material markets, and emerging business opportunities in material recovery [77, 78]. Social benefits include new job opportunities in the recycling sector, and support to local economies through material sourcing [78].

Table 8.3 Integration of the circular economy principles in building materials

Circular economy principle	Description	References
Design for durability and adaptability	Create materials and buildings that last longer and can be modified or reused instead of demolished	[71]
Use of recycled and renewable materials	Incorporate materials like reclaimed wood, recycled metals, or renewable resources such as bamboo	[72]
Modular and prefabricated construction	Build using off-site manufactured components that can be easily assembled, disassembled, and reused	[73]
Material recovery and recycling at end-of-life	Implement efficient deconstruction to separate and recycle materials instead of sending them to landfill	[74, 75]
Digital tracking and material passports	Use digital tools to track material composition and lifecycle, facilitating reuse and recycling efforts	[69, 71]

Challenges related to the circular economy including regulatory barriers, lack of standardized material data, the need for industry-wide collaboration, and increased initial costs. The challenges can be mitigated thought the promotion and education about circular economy principles, establishing a supportive regulatory framework and collaboration among stakeholders.

8.3.1 Sustainable Building Materials

Sustainable building materials such as bamboo, wood, Cross-Laminated Timber (CLT), recycled materials, Structural Insulated Panels (SIPs), and locally sourced materials demonstrate potential in reducing its environmental impact. Wood, when harvested responsibly from well-managed forests, is a renewable resource that stores carbon and provides natural insulation. Bamboo is valued for its rapid growth and renewability, making it a good candidate for traditional hardwoods [79]. CLT offers strong, lightweight panels made by layering wood, reducing the need for carbon-intensive materials like concrete and steel [80]. Incorporating recycled and reclaimed materials minimizes waste and lowers demand for extracted raw materials [81]. SIPs combine insulation and structural support in prefabricated panels, enhancing energy efficiency and reducing construction waste [82]. Additionally, using locally sourced materials reduces transportation emissions and supports regional economies.

Bamboo is among the most rapidly growing plants in the world [83]. It is an effective tool for carbon sequestration that offers new opportunities for sustainable livelihoods and new incomes in different communities and countries. Due to its resilience, it thrives on many types of soil, from poor to mineral-rich. It also survives extreme weather conditions such as drought, flooding and inundation. Bamboo can also withstand extreme temperatures from −20 °C to hot jungles. At the ecosystem level, bamboo plays an important role in repairing degraded landscapes, regulating erosion and protecting watersheds [84] In addition to its rapid growth and high renewable potential, bamboo matures within 2 to 5 years, making it an excellent choice for construction materials. Bamboo not only grows significantly faster than softwood or hardwood, it also releases 36% more oxygen into the atmosphere and stores significantly more CO_2 [85]. Bamboo's rapid growth provides enough raw materials to meet the ever-increasing demand for housing. Bamboo has strong fibers, and its compressive strength is bigger than that of concrete, while its tensile strength is almost equal to the steel strength [86]. Since bamboo materials can be recycled it is also responsible for less construction waste.

Cross-laminated timber (CLT) has emerged as an alternative in building construction due to its low carbon footprint, high strength-to-weight ratio, ease of installation, and aesthetic properties. CLT was invented in the early 1990s [87], made up of multiple layers of lumber bord stack and glued together to create a large scale, solid panel. As a building material, it possesses exceptional dimensional stability, high in-plane stiffness, and strength due to the way it is manufactured [63]. Due to its properties, there has been a surge of interest in the usage of CLT, in fact, a recent survey in the United States [88], which targeted the architectural community, found that more than half of participants considered CLT to be 'very suitable' as a primary construction material for multi-family residential buildings, while just under 5% said it was 'not suitable at all'.

SIPs are a panelized structural system composed of oriented strand board (OSB) wood panels glued to a lightweight board or foam core that is poured into place [89]. The use of SIPs in construction offers many environmental benefits such as reduced carbon emissions and improved thermal performance. Combined with renewable technologies such as solar photovoltaics, SIP homes can achieve near-zero energy building standards. Research from 2022 shows that a SIP home can meet all its energy requirements on-site. In this same study it is shown that by minimizing energy consumption and carbon emissions, SIPs construction supports the UK/EU targets to reduce carbon emissions by 2050 [90]. The other environmental advantage of SIPs is that they can be recycled and reused. This possibility of recycling minimizes waste and extends the lifespan of the materials and reduces costs.

8.4 The Path from Zero to Positive: A New Energy Paradigm

Achieving net-zero energy standards is not seen as the final goal, but a steppingstone towards a more ambitious vision, a net-positive performance. Net-positive performance is an emerging paradigm and represents buildings that generate surplus clean energy, rather than meeting only their own energy needs [91].

Net-Positive Buildings (*or Positive Energy Buildings*) are envisioned as the next step beyond the ZEB standard, delivering increased sustainability and long-term environmental benefits annually. Positive Energy Buildings (PEB) are a new concept, so there is no official definition [92]. However, as a part of Horizon 2020, the European Union has developed a PEB model. To achieve a status of PEB, many requirements must be fulfilled, such as occupant comfort, building efficiency and renewable energy provision [93]. Additionally, there are many studies that are investigating PEB and its definition [92]. One such project defined PEB as an energy-efficient building that generates more energy than it consumes using renewable sources. Indoor comfort is one of key aspects of PEBs [92, 94].

The challenge for evaluating net-positive building is that net-positive building goal goes beyond energy conservation. They apply unique technical, behavioral, political, and regulatory challenges that differ significantly from those of ZEB. They focus on requiring all new buildings to aim for net zero efficiency, but also on addressing the problem through individual buildings [95]. On the other hand, net-positive buildings goal is to explore the possibility of adding value for users and occupants, increased productivity and energy saving.

8.5 Conclusion

This chapter emphasizes current and future trends in zero energy buildings, through the discussion of technologies driving ZEBs. Advancements in building envelope design, smart HVAC systems, renewable energy integration, energy storage solutions, and artificial intelligence, this vision of ZEBs is becoming increasingly achievable. The transition to ZEBs faces certain challenges such as high initial costs, technological complexity, and maintenance. These challenges must be addressed through innovation, policy support, and greater awareness in the construction sector. The integration of circular economy principles in building materials further enhances the environmental and economic value of ZEBs.

The evolution from zero to positive energy buildings in the future offers the next frontier in sustainable architecture. This shift empowers buildings to serve as active contributors in advancing a cleaner and smarter energy grid.

References

1. Xu Y, Ramanathan V, Victor DG (2018) Global warming will happen faster than we think. Nature 564:30–32. https://doi.org/10.1038/D41586-018-07586-5
2. Ali G, Ee A, John's S, Newfoundland C (2017) Design and analysis of a net-zero energy house and its power system for Libya
3. Ban J, Xu D, He MZ, Sun Q, Chen C, Wang W, Zhu P, Li T (2017) The effect of high temperature on cause-specific mortality: a multi-county analysis in China. Environ Int 106:19. https://doi.org/10.1016/J.ENVINT.2017.05.019
4. Brown PT, Caldeira K (2017) Greater future global warming inferred from Earth's recent energy budget. Nature 552(7683):45–50. https://doi.org/10.1038/nature24672
5. Amro A. Building Europe's net-zero future why the transition to energy efficient and electrified buildings strengthens Europe's economy we are grateful to the following organisations for contributing their expertise and insight
6. Premrov M, Zigart M, Leskovar VZ (2017) Influence of the building geometry on energy efficiency of timber-glass buildings for different climatic regions. J Appl Eng Sci 15:529–539. https://doi.org/10.5937/jaes15-15256
7. Pajarskas K. Achieving a nearly zero-energy building (nZEB) status for a residential house in Finland
8. Campanella L (2024) Net zero energy buildings: an Italian case study. Analysis of the energy balance and retrofit hypothesis in order to reach the net zero energy target
9. Refaat SS et al (2021) Smart grid and enabling technologies. Wiley
10. Zhang M, Han W, He Y, Xiong J, Zhang Y (2024) Natural ventilation for cooling energy saving: typical case of public building design optimization in Guangzhou, China. Appl Sci 14:610. https://doi.org/10.3390/APP14020610
11. Jaffal I, Ouldboukhitine SE, Belarbi R (2012) A comprehensive study of the impact of green roofs on building energy performance. Renew Energy 43:157–164. https://doi.org/10.1016/J.RENENE.2011.12.004
12. Karimi M, Chikamoto T, Lee M, Tanaka T (2024) Impact of building orientation on energy performance of residential buildings in various cities across Afghanistan. Sustainability 16:11076. https://doi.org/10.3390/SU162411076
13. Sedaghatnia M, Faizi M, Khakzand M, Sanaieian H (2021) Energy and daylight optimization of shading devices, window size, and orientation for educational spaces in Tehran, Iran. J Architectural Eng 27:04021011. https://doi.org/10.1061/(ASCE)AE.1943-5568.0000466

14. Luible A, Gosztonyi S, Overend M, Aelenei L, Krstic-Furundzic A, Perino M, Goia F, Wellershoff F, Attia S, Pottgiesser U, Knaack U, Louter C. Facade 2018-adaptive! editors in chief: facade 2018-adaptive!
15. Gonçalves M, Figueiredo A, Almeida RMSF, Vicente R (2024) Dynamic façades in buildings: a systematic review across thermal comfort, energy efficiency and daylight performance. Renew Sustain Energy Rev 199:114474. https://doi.org/10.1016/J.RSER.2024.114474
16. Hosseini SM, Mohammadi M, Rosemann A, Schröder T, Lichtenberg J (2019) A morphological approach for kinetic façade design process to improve visual and thermal comfort: review. Build Environ 153:186–204. https://doi.org/10.1016/J.BUILDENV.2019.02.040
17. Iken O, Dlimi M, Agounoun R, Kadiri I, Fertahi SD, Zoubir A, Sbai K (2019) Numerical investigation of energy performance and cost analysis of Moroccan's building smart walls integrating vanadium dioxide. Solar Energy 179:249–263. https://doi.org/10.1016/J.SOLENER.2018.12.062
18. Shafaghat A, Keyvanfar A (2022) Dynamic façades design typologies, technologies, measurement techniques, and physical performances across thermal, optical, ventilation, and electricity generation outlooks. Renew Sustain Energy Rev 167:112647. https://doi.org/10.1016/J.RSER.2022.112647
19. Loonen RCGM, Rico-Martinez JM, Favoino F, Brzezicki M, Menezo C, La Ferla G, Aelenei L (Laura) (2015) Design for façade adaptability: Towards a unified and systematic characterization. https://research.tue.nl/en/publications/design-for-fa%C3%A7ade-adaptability-towards-a-unified-and-systematic-c
20. Borràs JG, Mas Á, Lerma C (2021) Green roof as a sustainable and energy efficient construction tool 13–27. https://doi.org/10.1007/978-3-030-71237-2_2
21. Kachenchart B, Panprayun G (2024) Selection of tropical plants for an extensive green roof with abilities of thermal performance, energy conservation, and greenhouse gas mitigation. Build Environ 265:112029. https://doi.org/10.1016/J.BUILDENV.2024.112029
22. Sirimanna MPG, Attalage RA (2016) A model for analyzing the thermal performance of roof configurations with flat inclined surfaces. Energy Build 116:122–132. https://doi.org/10.1016/J.ENBUILD.2015.12.052
23. Wu W, Skye HM, Domanski PA (2018) Selecting HVAC systems to achieve comfortable and cost-effective residential net-zero energy buildings. Appl Energy 212:577–591. https://doi.org/10.1016/J.APENERGY.2017.12.046
24. Wu W, Skye HM (2018) Net-zero nation: HVAC and PV systems for residential net-zero energy buildings across the United States. Energy Convers Manag 177:605–628. https://doi.org/10.1016/J.ENCONMAN.2018.09.084
25. Leng S, Canale L, Dupuis P, Zissis G (2016) Aging study of white high power LED under thermal and electrical stresses—new experimental setup prototype, photometric and electrical characterizations. Conference on lasers and electro-optics (2016), paper ATh3K.4. ATh3K.4. https://doi.org/10.1364/CLEO_AT.2016.ATH3K.4
26. Shankar A, Krishnasamy V, Chitti Babu B (2025) Smart LED lighting system with occupants' preference and daylight harvesting in office buildings. Energy Sources, Part A. https://doi.org/10.1080/15567036.2020.1859650
27. Obioma P, Agbodike O, Chen J, Wang L (2025) ISLS: IoT-based smart lighting system for improving energy conservation in office buildings
28. Wang N, Gao C, Ding C, Jia HZ, Sui GR, Gao XM (2019) A thermal management system to reuse thermal waste released by high-power light-emitting diodes. IEEE Trans Electron Devices 66:4790–4797. https://doi.org/10.1109/TED.2019.2938712
29. Li J, Zhang X, Zhou C, Zheng J, Ge D, Zhu W (2016) New applications of an automated system for high-power LEDs. IEEE/ASME Trans Mechatron 21:1035–1042. https://doi.org/10.1109/TMECH.2015.2487507
30. Kayode Olajiga O, Chigozie Ani E, Queen Sikhakane Z, Michael Olatunde T (2024) A comprehensive review of energy-efficient lighting technologies and trends. Eng Sci Technol J 5:1097–1111. https://doi.org/10.51594/estj/v5i3.973
31. Aussat Y. A power-efficient smart lighting system: modeling, implementation and evaluation

32. Pohl J, Frick V, Hoefner A, Santarius T, Finkbeiner M (2021) Environmental saving potentials of a smart home system from a life cycle perspective: how green is the smart home? J Clean Prod 312:127845. https://doi.org/10.1016/J.JCLEPRO.2021.127845
33. Lee S, Nengroo SH, Jin H, Doh Y, Lee C, Heo T, Har D (2022) Power management in smart residential building with deep learning model for occupancy detection by usage pattern of electric appliances
34. Selvaraj R, Kuthadi VM, Baskar S (2023) Smart building energy management and monitoring system based on artificial intelligence in smart city. Sustain Energy Technol Assess 56:103090. https://doi.org/10.1016/J.SETA.2023.103090
35. Wang Y, Hu B, Meng X, Xiao R (2024) A comprehensive review on technologies for achieving zero-energy buildings. Sustainability 16:10941. https://doi.org/10.3390/SU162410941
36. Sharma A (2024) Smart grid technologies for enhanced energy distribution. Darpan Int Res Anal 12:24–37. https://doi.org/10.36676/DIRA.V12.I3.54
37. Moura PS, López GL, Moreno JI, De Almeida AT (2013) The role of smart grids to foster energy efficiency. Energy Effic 6:621–639. https://doi.org/10.1007/s12053-013-9205-y
38. Javaid N, Ullah I, Akbar M, Iqbal Z, Khan FA, Alrajeh N, Alabed MS (2017) An intelligent load management system with renewable energy integration for smart homes. IEEE Access. 5:13587–13600. https://doi.org/10.1109/ACCESS.2017.2715225
39. Khalid M (2024) Smart grids and renewable energy systems: perspectives and grid integration challenges. Energ Strat Rev 51:101299. https://doi.org/10.1016/J.ESR.2024.101299
40. Giovanniello MA, Wu XY (2023) Hybrid lithium-ion battery and hydrogen energy storage systems for a wind-supplied microgrid. Appl Energy 345:121311. https://doi.org/10.1016/J.APENERGY.2023.121311
41. Flow batteries for grid-scale energy storage | MIT News | Massachusetts institute of technology. https://news.mit.edu/2023/flow-batteries-grid-scale-energy-storage-0407. Accessed 10 Jan 2025
42. Jafarizadeh H, Yamini E, Zolfaghari SM, Esmaeilion F, Assad MEH, Soltani M (2024) Navigating challenges in large-scale renewable energy storage: barriers, solutions, and innovations. https://doi.org/10.1016/j.egyr.2024.08.019
43. Maka AOM, Alabid JM (2022) Solar energy technology and its roles in sustainable development. Clean Energy 6:476–483. https://doi.org/10.1093/CE/ZKAC023
44. Belloni E, Bianchini G, Casini M, Faba A, Intravaia M, Laudani A, Lozito GM (2024) An overview on building-integrated photovoltaics: technological solutions, modeling, and control. Energy Build 324:114867. https://doi.org/10.1016/J.ENBUILD.2024.114867
45. Ghosh A, Sarmah N, Sundaram S, Mallick TK (2019) Numerical studies of thermal comfort for semi-transparent building integrated photovoltaic (BIPV)-vacuum glazing system. Sol Energy 190:608–616. https://doi.org/10.1016/J.SOLENER.2019.08.049
46. Chandel SS, Gupta A, Chandel R, Tajjour S (2023) Review of deep learning techniques for power generation prediction of industrial solar photovoltaic plants. Solar Compass 8:100061. https://doi.org/10.1016/J.SOLCOM.2023.100061
47. Zhao H, Yang RJ, Liu C, Sun C (2023) Solar building envelope potential in urban environments: a state-of-the-art review of assessment methods and framework. Build Environ 244:110831. https://doi.org/10.1016/J.BUILDENV.2023.110831
48. ALRubeei IRN, Idi SN, Hussein Alsammak IL, AlRikabi HT, Mutar HA, Alaidi AHM (2025) Using artificial intelligence for enhancement of solar cell efficiency in the of Iraq. Heritage Sustain Dev 7:197–210. https://doi.org/10.37868/HSD.V7I1.1067
49. Das Karmakar S, Chattopadhyay H (2022) A review of augmentation methods to enhance the performance of vertical axis wind turbine. Sustain Energy Technol Assess 53:102469. https://doi.org/10.1016/J.SETA.2022.102469
50. Ghamari M, Sundaram S (2024) Solar window innovations: enhancing building performance through advanced technologies. https://doi.org/10.3390/en17143369
51. González Gallero FJ, González Siles G, Rodríguez Maestre I, Foncubierta Blázquez JL, Bottarelli M (2024) Performance analysis of a novel façade-based building integrated photovoltaic-thermal system with phase change material for domestic hot water. Appl Therm Eng 240:122321. https://doi.org/10.1016/J.APPLTHERMALENG.2023.122321

52. Badea A, Baracu T, Dinca C, Tutica D, Grigore R, Anastasiu M (2014) A life-cycle cost analysis of the passive house Politehnica from Bucharest. Energy Build 80:542–555. https://doi.org/10.1016/J.ENBUILD.2014.04.044
53. (PDF) Artificial intelligence and life in 2030: the one hundred year study on artificial intelligence. https://www.researchgate.net/publication/365359355_Artificial_Intelligence_and_Life_in_2030_The_One_Hundred_Year_Study_on_Artificial_Intelligence. Accessed 10 Jan 2025
54. Ng K, Ng KW. Big data: understanding and implementation in the data-driven era
55. Heti K, Silvanskaya G, Khambir V, Harkava V, Ivanchenko K (2025) The analysis of engineering adaptation to global challenges and strategic planning for the future. Sustain Eng Innov 7:137–146. https://doi.org/10.37868/SEI.V7I1.ID453
56. Yesilyurt H, Dokuz Y, Dokuz AS (2024) Data-driven energy consumption prediction of a university office building using machine learning algorithms. Energy 310:133242. https://doi.org/10.1016/J.ENERGY.2024.133242
57. Wang Y, Chen Q, Hong T, Kang C (2019) Review of smart meter data analytics: applications, methodologies, and challenges. IEEE Trans Smart Grid 10:3125–3148. https://doi.org/10.1109/TSG.2018.2818167
58. Wen L, Zhou K, Yang S, Li L (2018) Compression of smart meter big data: a survey. Renew Sustain Energy Rev 91:59–69. https://doi.org/10.1016/J.RSER.2018.03.088
59. Wuyun Q, Li B, Bian M, Wang C, Huang Z, Wang B, Cai W, Wang M, Zhang X, He T, Xu W (2024) Demonstration and data analysis of a zero emission building (ZEB) in Beijing, China. Sol Energy 272:112488. https://doi.org/10.1016/J.SOLENER.2024.112488
60. Mahesh B (2020) Machine learning algorithms—a review. Int J Sci Res (IJSR) 9:381–386. https://doi.org/10.21275/art20203995
61. Assareh E, Kazemiani-Najafabadi P, Amiri Rad E, Firoozzadeh M, Farhadi E, Agarwal S, Wang X, Hosseinzadeh M, Pak W (2024) Machine learning optimization of a zero energy building (ZEB) by waste heat recovery in a co-generation system. Energy Build 323:114711. https://doi.org/10.1016/J.ENBUILD.2024.114711
62. Siasiev A, Dudnikov S, Nikolaienko D, Hubal H, Bondarchuk O (2025) Evaluation of the effectiveness of decision-making algorithms in the management of renewable energy sources: a case study of Ukraine. Sustain Eng Innov 7:87–96. https://doi.org/10.37868/SEI.V7I1.ID415
63. Onwusinkwue S, Osasona F, Ahmad IAI, Anyanwu AC, Dawodu SO, Obi OC, Hamdan A (2024) Artificial intelligence (AI) in renewable energy: a review of predictive maintenance and energy optimization. World J Adv Res Rev 21:2487–2799. https://doi.org/10.30574/wjarr.2024.21.1.0347
64. Al-Emran M (2023) Beyond technology acceptance: development and evaluation of technology-environmental, economic, and social sustainability theory. Technol Soc 75:102383. https://doi.org/10.1016/J.TECHSOC.2023.102383
65. Huo D, Gu W, Guo D, Tang A (2024) The service trade with AI and energy efficiency: multiplier effect of the digital economy in a green city by using quantum computation based on QUBO modeling. Energy Econ 140:107976. https://doi.org/10.1016/J.ENECO.2024.107976
66. Năstasă A, Dumitra TC, Grigorescu A (2024) Artificial intelligence and sustainable development during the pandemic: an overview of the scientific debates. Heliyon 10:e30412. https://doi.org/10.1016/J.HELIYON.2024.E30412
67. Zheng X, Ma Z, Yuang Z (2024) Urban design and pollution using AI: implications for urban development in China. Heliyon 10:e37735. https://doi.org/10.1016/J.HELIYON.2024.E37735
68. Song K, Jang Y, Park M, Lee HS, Ahn J (2020) Energy efficiency of end-user groups for personalized HVAC control in multi-zone buildings. Energy 206:118116. https://doi.org/10.1016/J.ENERGY.2020.118116
69. Charef R (2024) A digital framework for the implementation of the circular economy in the construction sector: expert opinions. Sustainability16:5849. https://doi.org/10.3390/SU16145849
70. Shaznie A, Sinaau A, Waheeda A, Muna F, Chee Hoo W (2025) Unlocking sustainability: exploring the benefits and challenges of implementing circular economy principles in the Maldives. Heritage Sustain Dev 7:355–370. https://doi.org/10.37868/HSD.V7I1.872

71. Pomponi F, Moncaster A (2017) Circular economy for the built environment: a research framework. J Clean Prod 143:710–718. https://doi.org/10.1016/J.JCLEPRO.2016.12.055
72. Implementing circular economy principles in construction. https://instituteofsustainabilitystudies.com/insights/guides/implementing-circular-economy-principles-in-construction/?utm_source=chatgpt.com. Accessed 27 June 2025
73. Huang R, Yu J, Wang S, Ottelin J, Cetinay H, Behrens P, Mackenbach S, Zeller JC, Osebold R (2020) A roadmap towards circularity—modular construction as a tool for circular economy in the built environment. IOP Conf Ser Earth Environ Sci 588:052027. https://doi.org/10.1088/1755-1315/588/5/052027
74. Minunno R, O'Grady T, Morrison GM, Gruner RL, Colling M (2018) Strategies for applying the circular economy to prefabricated buildings. Buildings 8:125. https://doi.org/10.3390/BUILDINGS8090125
75. Ding GKC (2008) Sustainable construction—the role of environmental assessment tools. J Environ Manage 86:451–464. https://doi.org/10.1016/J.JENVMAN.2006.12.025
76. Firoozi AA, Firoozi AA (2022) Circular economy for sustainable construction material management. J Civil Eng Urbanism 12:70–81. https://doi.org/10.54203/JCEU.2022.10
77. Finamore M, Oltean-Dumbrava C (2025) Emerging trends in the circular economy: multidimensional perspective in the building sector. Circular Econ Sustain 1–36. https://doi.org/10.1007/S43615-024-00485-0/TABLES/5
78. Valencia M, Bocken N, Loaiza C, De Jaeger S (2023) The social contribution of the circular economy. J Clean Prod 408:137082. https://doi.org/10.1016/J.JCLEPRO.2023.137082
79. Adier MFV, Sevilla MEP, Valerio DNR, Ongpeng JMC (2023) Bamboo as sustainable building materials: a systematic review of properties, treatment methods, and standards. Buildings 13:2449. https://doi.org/10.3390/BUILDINGS13102449
80. Hossain MU, Ng ST, Antwi-Afari P, Amor B (2020) Circular economy and the construction industry: existing trends, challenges and prospective framework for sustainable construction. Renew Sustain Energy Rev 130:109948. https://doi.org/10.1016/J.RSER.2020.109948
81. Buratti C, Belloni E, Lascaro E, Lopez GA, Ricciardi P (2016) Sustainable panels with recycled materials for building applications: environmental and acoustic characterization. Energy Procedia 101:972–979. https://doi.org/10.1016/J.EGYPRO.2016.11.123
82. Ouakarrouch M, Bousshine S, Bybi A, Laaroussi N, Garoum M (2022) Acoustic and thermal performances assessment of sustainable insulation panels made from cardboard waste and natural fibers. Appl Acoust 199:109007. https://doi.org/10.1016/J.APACOUST.2022.109007
83. Fei B, Gao Z, Wang J, Liu Z (2016) Biological, anatomical, and chemical characteristics of Bamboo. Secondary Xylem Biol: Origins, Functions, Appl 283–306. https://doi.org/10.1016/B978-0-12-802185-9.00014-0
84. Tripathi A, Yadav S, Nishtha, Nkengnamai M, Thakur A (2024) Bamboo: a fast-growing species to mitigate carbon footprint. For Clim Change 469–487. https://doi.org/10.1007/978-981-97-3905-9_23
85. Muleya F, Mwanaumo E, Musondo I (2018) DII-2018: 5th international conference on development and investment in infrastructure strategies for Africa: 11–13 July 2018, Livingstone, Zambia. The development and investment in infrastructure (DII) conference series
86. Yadav M, Mathur A (2021) Bamboo as a sustainable material in the construction industry: an overview. Mater Today Proc 43:2872–2876. https://doi.org/10.1016/J.MATPR.2021.01.125
87. Younis A, Dodoo A (2022) Cross-laminated timber for building construction: a life-cycle-assessment overview. J Build Eng 52:104482. https://doi.org/10.1016/J.JOBE.2022.104482
88. Laguarda Mallo MF, Espinoza O (2015) Awareness, perceptions and willingness to adopt cross-laminated timber by the architecture community in the United States. J Clean Prod 94:198–210. https://doi.org/10.1016/J.JCLEPRO.2015.01.090
89. Jacques E, Makar J (2019) Behaviour of structural insulated panels (SIPs) subjected to short-term out-of-plane transverse loads. Can J Civ Eng 46:858–869. https://doi.org/10.1139/CJCE-2018-0015
90. Finnegan S, Edwards R, Al-Derbi B, Campbell I, Fulton M (2022) The potential of structurally insulated panels (SIPs) to supply net zero carbon housing. Buildings 12. https://doi.org/10.3390/buildings12122081

91. Agharid AP, Permana I, Wang F, Lee RC (2025) Energy performance and decarbonization evaluation of a novel positive energy building using solar PV. Build Environ 269:112397. https://doi.org/10.1016/J.BUILDENV.2024.112397
92. Ala-juusela M, Rehman HU, Hukkalainen M, Reda F (2021) Positive energy building definition with the framework, elements and challenges of the concept. Energies (Basel) 14. https://doi.org/10.3390/EN14196260
93. Mousavi SN, Villarreal-Marroquín MG, Hajiaghaei-Keshteli M, Smith NR (2023) Data-driven prediction and optimization toward net-zero and positive-energy buildings: a systematic review. Build Environ 242:110578. https://doi.org/10.1016/J.BUILDENV.2023.110578
94. Petrunenko I, Zhuk O, Litvak O, Litvak S, Yemets V (2024) The impact of renewable energy sources on economic recovery in Ukraine. Sustain Eng Innov 6:297–308. https://doi.org/10.37868/SEI.V6I2.ID399
95. Cole RJ, Fedoruk L (2015) Shifting from net-zero to net-positive energy buildings. Build Res Inf 43:111–120. https://doi.org/10.1080/09613218.2014.950452

MIX
Papier aus verantwortungsvollen Quellen
Paper from responsible sources
FSC® C105338

If you have any concerns about our products,
you can contact us on
ProductSafety@springernature.com

In case Publisher is established outside the EU,
the EU authorized representative is:
Springer Nature Customer Service Center GmbH
Europaplatz 3, 69115 Heidelberg, Germany

Printed by Libri Plureos GmbH
in Hamburg, Germany